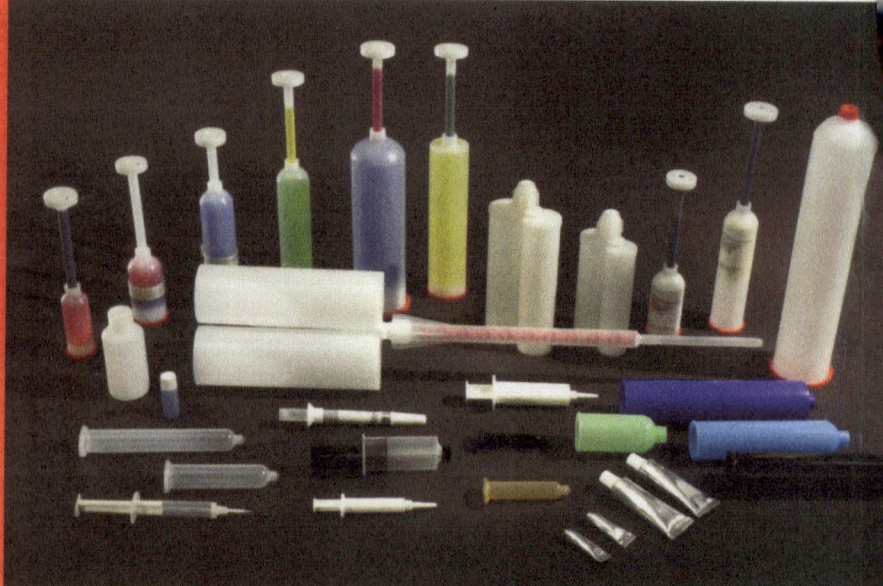

ADHESIVE TECHNOLOGY COMPENDIUM 2009

adhesion ADHESIVES & SEALANTS
Industrieverband Klebstoffe e. V.
(German Adhesives Association)

VIEWEG+
TEUBNER

Editor: adhesion ADHESIVES & SEALANTS
                   Abraham-Lincoln-Straße 46
                   D-65189 Wiesbaden
                   Phone +49 (0) 6 11-78 78-2 83
                   Fax +49 (0) 611-78 78-4 95
                   www.adhaesion.com
                   Email: marlene.doobe@gwv-fachverlage.de

                   Supported by:
                   Industrieverband Klebstoffe e. V.
                   German Adhesives Association
                   Völklinger Straße 4 (RWI-Haus)
                   D-40219 Düsseldorf
                   Phone +49 (0) 2 11-6 79 31 10
                   Fax +49 (0) 2 11-6 79 31 33

Publishing house: Vieweg+Teubner | GWV Fachverlage GmbH
                   Abraham-Lincoln-Straße 46
                   D-65189 Wiesbaden
                   www.viewegteubner.de

All company profiles and sources are based on the information provided by the respective companies (as at July 2009). The publishing house does not accept any responsibility for the completeness and correctness of the information.

Vieweg+Teubner is a member of the Springer Science+Business Media group.

The use of addresses and other information are a breach of the competition law and copyright law and are subject to legal action.

Print: Krips b. v., Meppel

Layout: Satzwerk GbR, D-63303 Dreieich

ISBN 978-3-8348-0979-7   Nominal sum: € 25.90

Dear Reader

Adhesion is by far the most important technology of the 21st century – today there is hardly any industry that does not bank on the use of this innovative and dependable joining technology.

In the key disciplines of automotive engineering, aerospace and electrical engineering, impressive developments have taken place over recent decades. Today it seems completely matter-of-course that we drive crash-optimized cars, fly around the globe and use ID cards with microchips that contain biometric data. The development of adhesives over the same period is no less impressive. Yesterday's handy home help has turned into a truly high tech product. From airplanes to zoom lenses, advanced adhesives and sealants are the very enablers of modern-day lifestyles.

The German adhesives and sealants industry holds a special position: It is considered a technology leader both in Europe and globally. Adhesives and sealants "Made in Germany" enjoy increasing popularity and are used worldwide; in fact one third of the products manufactured in Germany are destined for export markets.

Every year, over 900,000 tonnes of adhesives and sealants and more than 800 million square metres of adhesive tape are produced in Germany. Total sales across the industry, including adhesives, sealants, adhesive tape and cement-based adhesives for use in the construction industry add up to some three billion Euros.

Indeed the cutting edge German adhesives industry makes quite a contribution to the national economy: The use of adhesives and sealants by almost all industries creates potential value in excess of 330 billion Euros.

The German Adhesives Association currently represents the economic and technological interests of 122 manufacturers of adhesive, raw materials, adhesive tape and sealants.

From our perspective, one of our key tasks is to keep the public informed about this innovative joining technology, its manufacturers and our association on a regular basis.

We are therefore delighted to present the latest edition of our Adhesive Technology Compendium which we have prepared jointly with the editor "adhesion Kleben & Dichten". It contains interesting facts about the adhesives industry and our association, documents the profiles of member companies as well as major system partners and scientific institutes.

With this 8th edition of our compendium, we provide a helpful tool for day to day work and research in the field of adhesives and sealants for all those who are interested in adhesion and sealing technology.

Dr. Ralf Schelbach
Chairman
Industrieverband Klebstoffe e. V.

Ansgar van Halteren
Senior Executive
Industrieverband Klebstoffe e. V.

German
Adhesives
Association
Industrieverband Klebstoffe e.V.

adhesion ADHESIVES & SEALANTS

MADE
IN
GERMANY

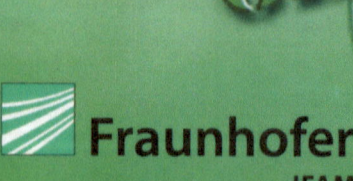
Fraunhofer

IFAM

EMICODE® GEV

Industrieverband Klebstoffe e. V.
Völklinger Straße 4 (RWI-Haus)
D-40219 Düsseldorf
Phone +49 (0) 2 11-6 79 31 10, fax +49 (0) 2 11-6 79 31 33

Industrieverband
Klebstoffe e.V.

List of Advertisers

Cover: Fraunhofer Institute for Manufacturing Technology and Applied Materials Research (IFAM)
Advice: Due to technical reasons it was not always possible to keep the alphabetic order of the company profiles.

Adhesive Technology – the Key to Innovative Products
Unlimited Variety of Applications

In industrial applications, bonding has long proved to be at least as good as – and in some cases even superior to – conventional joining technologies including welding, soldering, riveting and fastening with bolts. This advanced technology today enjoys the reputation of a problem solver which helps meet ever more stringent customer demands.

The significance of adhesive bonding technology is amply demonstrated by the example of environmental protection. In addition to the issues involved with climate change, greenhouse gases and CO_2 emissions, cost is becoming an increasingly important factor. As is widely known, the increase in the price of energy from fossil fuels including coal, oil and natural gas have obliged us to look for alternative sources of energy. In Germany, 20,000 wind turbines have been installed, and people building new houses often put solar panels on the roof right from the start.

The Statistics are promising: According to the German Federal Association of the Energy and Water Industry (BDEW), as much as 14 percent of total power consumption in Germany already comes from renewable energy sources and the trend is set to increase. For both wind or solar power, adhesives help ensure an uninterrupted supply of green power. When it comes to the construction of these plants, conventional connecting techniques such as welding are at their limit: after all, solar plants and wind turbines must withstand 20 to 30 years of extreme climatic conditions including wind and snow, radiation and erosion and major temperature fluctuations. What is more, wind turbines are sometimes subject to high wind speeds. And they are huge: the rotor blades can be up to 50 metres long, which leads to high loads from aerodynamics and dead weight. That said, gluing is much safer than a number of alternative techniques.

Modern rotor blades are glued. That makes the use of wind power safer and more economically efficient. (Picture credits: obs/DNR/ Herbert Grabe).

When bonded, materials are not weakened by punched or drilled holes, and no unwanted tension is produced by soldered or welded spots. The forces acting upon adhesives are transferred across a large area, and there is no risk of corrosion.

For example, reactive adhesives such as epoxy resin are used in the production of rotor blades. These are first tested for tensile strength and elongation at break based on various safety regulations. That way, it can be ensured that glued connections will reliably withstand wind and weather.

Mobile telephony – impossible without adhesives

Mobile telephony is another prominent example of the huge range of possibilities that can be realized with adhesives. Over the past few years, mobile phones were able to become so small and cheap thanks to adhesive technology. Since mobile telephony first came into being, the technology has developed at breakneck speed. Today's all-rounders have little in common with the mobile phones of the 1980s.

When buying a mobile phone, it's all about finding a "fancy, stylish and functional" handheld. In the 21st century, it goes without saying that this intelligent aid fits into any jacket pocket. And prices have dropped too. With WAP, Bluetooth or GPRS – advanced mobile phones have become so affordable that some people have more than just one.

Smaller, better, cheaper: adhesives and adhesive foils have played an important part in this almost incredible success story.

To give an example: For the current to flow in a mobile phone, various contact points must be connected. Advanced electrically conducting adhesive films based on epoxy resin, silicon or

Never without a mobile phone. Adhesives have enabled the development of small, cheap mobile communication gadgets, allowing you to call the breakdown service for instance.

polyurethane have replaced conventional bolted or soldered connections. This means that the distance between contacts can now be as little as 0.035 mm.

Thus electric connections have become slimmer, less complex and cheaper, all thanks to adhesives. *Better* and *safer* are other attributes of adhesive films: They do not become as hot as soldered connections, and they are free from heavy metals. In addition, up to 21 different adhesive foils in a mobile phone protect users from voltage and prevent moisture from penetrating. Finally, it is adhesive foils that protect the display from scratches.

Lightweight cars with polymer foam

In the car industry, improvements to meet more demanding customer requirements could hardly be realized without modern adhesive technology. Lighter, faster and safer is what it's all about. These trends call for polymer foams that allow thick sheets to be replaced by thin-walled hollow components. The filler is a special plastic and a special substance that makes the blend expand as it is heated. The result: lightweight foam components which provide high protection in the event of a crash. From a small second car to a comfortable saloon car, adhesives give that extra margin of safety. For some car body components, adhesives have already been the preferred material for some 40 years now: Adhesive bonded connections offer a range of benefits including homogeneous stress and force transmission. In addition, they allow different materials to be connected without impairing the material properties of the parts joined together.

Thanks to innovative, high performance adhesives, connections have become possible in car manufacturing which a few years ago would have been unthinkable: glass and steel, aluminium and magnesium, fibre composites and metal. In addition, adhesives help integrate additional functionality such as vibration absorption, electric insulation or protection against corrosion. In order to reduce petrol consumption and emissions, car manufacturers and suppliers are moving

Saving petrol and reducing emissions - using adhesives, lightweight constructions have revolutionized the automotive industry.

increasingly towards lightweight designs. With car weight as the determining factor future improvements will come through innovative concepts like polymer foams. Research departments are working at full steam – for instance on a second lightweight alternative: metal foams. These are made up of small pellets that are enveloped with adhesive using a special process. This mass is filled into hollows where the pellets bond together. Metal foam guarantees significantly lower weight whilst offering convincing crash resistance.

Adhesives in aerospace

It is impossible to imagine aerospace without innovative bonds, which must meet requirements more extreme than those of any other industry. For cost and weight reasons and with a view to maximum safety, many spacecraft parts are bonded. The heat shield from ceramic tiles is one of the most heavily loaded areas: Within seconds the temperature in space can fluctuate by several hundred degrees Celsius, for instance when a spacecraft emerges from the shadow of a planet into full sunlight.

What is more, minuscule dust particles in space should not be underestimated as they can hit the spacecraft so fast that they may leave cracks or small craters. European scientists from Bristol University in the UK look to have made a quantum leap in this area: on the behalf of the European Space Agency (ESA) they have developed an adhesive-containing material that repairs itself automatically in the event of damage. At first glance, however, this idea seems paradoxical.

An adhesive which contains hundreds of glass fibres is designed to reinforce material under even the most extreme conditions and that is what is so clever. These glass fibres are expected to burst once the outer shell of the spacecraft is damaged. Half of the fibres are filled with a resin-type adhesive while the other half contains a chemical substance. Thus when the glass fibres burst, the adhesive and chemical substance react together. The result: the material hardens immediately without the liquid adhesive evaporating in the vacuum. Thus a protective layer is formed, like encrusted blood on a wound. While this self-repairing material is expected to be ready for use in 2011 at the earliest, US researchers are also experimenting with self-healing materials. For instance, scientists from the University of Illinois are working on integrating small capsules into the material instead of glass fibres. But no matter how, when and where – what counts in the end is that sophisticated adhesives will ensure in the near (or distant) future that astronauts need not undertake the risky repair of the outer shell on their own.

Conclusion

In short, adhesive technology is indispensable when it comes to combining different materials while sustainably maintaining their properties. Only with adhesives can new engineering approaches be exploited. Furthermore, additional features can be integrated into the component which go beyond the mere bond, for instance vibration absorption, sealing against gas and liquids, compensation of different component dynamics, and protection against corrosion. Adhesive bonding goes beyond all other ways of connecting items together and enables advanced designs to be implemented by optimizing the combination of technological, economic and ecological factors. Adhesive technology is used across the whole A to Z of industries to produce everything from airplanes to zoom lenses and is ever-more important to industry and small trade. It offers enormous potential for innovation.

Many spacecraft parts are glued for reasons of safety and costs (Picture credits: ESA AOES Medialab)

Adhesive Producer
Raw Material Supplier

AB-Tec GmbH & Co. KG

Pütterstraße 22
D-58636 Iserlohn
Phone +49 (0) 23 71-94 91-25
Fax +49 (0) 23 71-46 01 93
Email: info@ab-tec.com
www.ab.tec.com

Company

Year of formation
2.000

Size of workforce
350 in the group

Sales channels
direct sales and distribution partners
throughout Europe

Contact partners
Management:
Kenneth Keuchel, Managing Director
Wolfgang Schulte, Managing Director

Application technology and sales:
Bernhard Kaschner, Sales Director

Further information
Thermoplastic heat-activatable adhesives,
dry, in roll-form, large widths, weights from
6 – 60 g/m²

Range of Products

Types of adhesives
Hot melt adhesives

Raw materials
Polymers:
Co-polymamide,
Co-polyester,
Co-polyolefine,
Polyurethane

For applications in the field of
Paper/packaging
Wood/furniture industry
Construction industry, including floors,
walls and ceilings
Mechanical engineering and equipment
construction
Automotive industry, aviation industry
Textile industry
Adhesive tapes, labels
Composites
Filters

Adtracon GmbH
Düsseldorfer Straße 96 – 100
D-40721 Hilden
Phone +49 (0) 21 03-7 73 58
Fax +49 (0) 21 03-7 73 59
Email: info@adtracon.de
www.adtracon.de

Member of IVK

Company

Year of formation
2003

Size of workforce
10

Managing partners
Dr. Roland Heider

Nominal capital
37.750 €

Ownership structure
Dr. Roland Heider, ISB, Kfw

Sales channels
Direct and distributors

Contact partners
Management:
Dr. Roland Heider

Further information
We focas on reactive hotmelts

Range of Products

Types of adhesives
Hot melt adhesives
Reactive adhesives

Types of sealants
PUR sealants

Raw materials
Additives
Fillers
Resins
Polymers

Equipment, plant and components
for conveying, mixing, metering and for
adhesive application

For applications in the field of
Bookbinding/graphic design
Wood/furniture industry
Automotive industry, aviation industry
Textile industry

Alberdingk Boley
Düsseldorfer Straße 53
D-47829 Krefeld
Phone +49 (0) 21 51-5 28-0
Email: info@alberdingk-boley.de
www.alberdingk-boley.de

Member of IVK

Company

Year of formation
1828

Size of workforce
240

Ownership structure
family-owned

Subsidiaries
Alberdingk Boley, Inc., Greensboro, USA,
Alberdingk Resins (Shenzhen) Co. Ltd.,
Shenzhen, China,
Thai Castor Oil Industries Co., Ltd.,
Bangkok, Thailand

Sales channels
worldwide

Contact partners
Management:
Dr. Gerhardt Leinz,
Research and Development

Application technology and sales:
Markus Dimmes, Manager Technical
Marketing Coatings
Johannes Leibl, Manager Sales Dispersions

Range of Products

Types of adhesives
Reactive adhesives
Dispersion adhesives
Pressure-sensitive adhesives

Raw materials
Resins
Polymers

For applications in the field of
Paper/packaging
Bookbinding/graphic design
Wood/furniture industry
Construction industry, including floors,
walls and ceilings
Electronics
Mechanical engineering and equipment
construction
Automotive industry, aviation industry
Textile industry
Adhesive tapes, labels
Hygiene
Household, recreation and office

ARDEX GmbH

Friedrich-Ebert-Straße 45
D-58453 Witten
Phone +49 (0) 2 30 26 64-0
Fax +49 (0) 2 30 26 64-7 37
Email: info@ardex.de
www.ardex.de

Member of IVK

Company

Year of formation
1949

Size of workforce
2.000

Ownership structure
Private

Subsidiaries
USA, UK, Denmark, Austria, Spain, Singapore,
India, China, Taiwan, Australia

Contact partners
Management:
Dieter A. Gundlach

Application technology and sales:
Daniel Händle

Range of Products

Types of adhesives
Dispersion adhesives
Pressure-sensitive adhesives

Types of sealants
Acrylic sealants
Butyl sealants
PUR sealants
Silicone sealants
MS / SMP sealants

For applications in the field of
Construction industry, including floors,
walls and ceilings

The Chemical Company

BASF SE
D-67056 Ludwigshafen
Phone +49 (0) 6 21-60-0
Email: industrial-adhesives@basf.com
pressure-sensitive-adhesives@basf.com
info-pib@basf.com
www.basf.com

Member of IVK

Company

Year of formation
1865

Size of workforce
~ 97.000 (year end 2008)

Further information
Areas of adhesive technology:
Pressure-sensitive adhesives
Industrial adhesives
Polyisobutene PIB

Range of Products

Raw materials for
Hot melt adhesives
Dispersion adhesives
Pressure-sensitive adhesives
Acrylic sealants
Butyl sealants
Additives
Resins
Polymers

For applications in the field of
Paper/packaging
Bookbinding/graphic design
Wood/furniture industry
Construction industry, including floors,
walls and ceilings
Mechanical engineering and equipment
construction
Automotive industry, aviation industry
Textile industry
Adhesive tapes, labels
Hygiene
Household, recreation and office

Bayer MaterialScience AG
D-51368 Leverkusen
Phone +49 (0) 2 14-30-7 18 41
Fax +49 (0) 2 14-30-96 7 18 41
Email: cas@bayermaterialscience.com
www.bayercoatings.com

Member of IVK

Company

Year of formation
2004

Size of workforce
15,100

Contact partners
Marketing Europe/Business Development
Phone +49 (0) 2 14-30-7 18 41
Fax +49 (0) 2 14-30-96 7 18 41
Email: cas@bayermaterialscience.com

Range of Products

Raw materials
Polymers:
Polyurethane–Dispersions (Dispercoll® U)
Hydroxylpolyurethanes (Desmocoll®,
Desmomelt®)
Polyisocyanates (Desmodur®)
Isocyanate-Prepolymers (Desmodur®,
Desmoseal®)
Silanterminated Polyurethanes
(Desmoseal® S)
Polyesterpolyols (Baycoll®)
Polyetherpolyols (Desmophen®, Acclaim®)
Polychloroprene-Dispersions (Dispercoll® C)
Polychloroprenes (Baypren®)
Halogenated Polyisoprenes (Pergut®)
Silicon dioxide-nanoparticle dispersions
(Dispercoll® S)

Berger-Seidle Siegeltechnik GmbH

Maybachstraße 2
D-67269 Grünstadt/Weinstraße
Phone +49 (0) 63 59-80 05-0
Fax +49 (0) 63 59-80 05-50
Email: info@berger-seidle.de
www.berger-seidle.de

Member of IVK

Company

Year of formation
1926

Size of workforce
85

Ownership structure
100 % subsidiary to Phil. Berger GmbH

Sales channels
Distributors and sales partners/representatives on each country

Contact partners
Management:
Markus M. Adam

Application technology and sales:
Wolfgang Brutzke

Further information
www.berger-seidle.de

Types of adhesives
Reactive adhesives
Solvent-based adhesives
Dispersion adhesives

Range of Products

Types of sealants
Acrylic sealants
PUR sealants
MS/SMP sealants

For applications in the field of
Wood/furniture industry
Construction industry, including floors, walls and ceilings

Bona AB

Murmansgatan 130, Box 210 74
S-20021 Malmö
Phone +46 40 38 55 00
Fax +46 40 18 16 43
Email: bona@bona.com

Member of IVK

Company

Year of formation
1909

Size of workforce
500

Ownership structure
privat owned stock cooperation

Subsidiaries
Asia Pacific: BonaKemi Asia Pacific Pte Ltd
Phone +65 6738 55 80
Austria: BonaKemi GesmbH
Phone +43 662 66 19 43-9
Belgium: Bona NV
Phone +32 2 721 2759
China: Bona Coatings (Dalian) Co., Ltd.
Phone +86 411 87585500
Czech Republic: Bona CR s.r.o.
Phone +420 236 080 211
France: Bona S.á.R.L.
Phone: +33 3 88 49 18 60
Germany (does include the production sites
for adhesives)
Bona GmbH Deutschland
Phone +49 6431 4008 0
Hungary: Bona Magyyarország Kft.
Phone +36 1 204 5617
Italy: Bona S.r.l.
Phone +39 045 8753 318
Latin America: Bona Brazil
Phone +55 47 9931 9525
Netherlands: Bona Benelux BV
Phone +31 23 542 1864,
Poland: Bona-Polska Sp. Z.o.o.
Phone +48 61 816 34 60

Range of Products

Types of adhesives
Reactive adhesives
Solvent-based adhesives
Dispersion adhesives

For applications in the field of
Construction industry, including floors,
walls and ceilings

Romania: Bona Romania S.r.l.
Phone +40 31 405 75, +40 21 317 02 22
Shanghai: Bona trading (Shanghai) Co., Ltd
Phone +86 21 62 36 1540/41
Spain: Bona Ibérica S.L
Phone +34 91 642 0188
United Kingdom: Bona Limited
Phone +44 1908 399 740
United States: BonaKemi USA, Inc
Phone +1 303 371 1411,
+1 800 872 5515 or +1 800 984 3871 toll
free in US and Canada

Contact partners
Management:
Lars Olle Andersson (sales)
Thomas Brokamp (production, R & D),
and the managers of the local subsidiaries

Application technology and sales:
Thomas Haack (technical)
Ferenc Barta (sales)

Further information
Have a look on our website: www.bona.com

Bostik GmbH
An der Bundesstraße 16
D-33829 Borgholzhausen
Phone +49 (0) 54 25-8 01-0
Email: info.germany@bostik.com
www.bostik.com

Member of IVK

Company

Year of formation
1903

Size of workforce
300

Nominal capital
3.000.000

Subsidiaries
MEM Bauchemie GmbH, Bostik Poland,
Bostik Hungary

Sales channels
Construction distribution,
industry

Contact partners
Management:
Hans Deike, Managing Director
Norbert Uniatowsky, division industry
Ulrich Springer, division construction
Manfred Nolting, finance & administration
Dr. Michael Müller, R & D
Bernd Köhler, production
Gerhard Flottmann, purchase
Dr. Hans-Ulrich Hürter, division transport
Olaf Memmen, MEM Bauchemie

Range of Products

Types of adhesives
Hot melt adhesives
Reactive adhesives
Solvent-based adhesives
Dispersion adhesives
Vegetable adhesives, dextrin and
starch adhesives
Pressure-sensitive adhesives

Types of sealants
Acrylic sealants
Butyl sealants
PUR sealants
Silicone sealants
MS/SMP sealants

Raw materials
Additives, Fillers, Resins, Solvents
Polymers, Starch

For applications in the field of
Paper/packaging
Bookbinding/graphic design
Wood/furniture industry
Construction industry, including floors,
walls and ceilings
Mechanical engineering and equipment
construction
Automotive industry, aviation industry
Textile industry
Adhesive tapes, labels
Hygiene
Household, recreation and office

Friedrich Branding GmbH & Co. KG

Postfach 1460
D-31254 Lehrte
Phone +49 (0) 51 32-50 08-0
Fax +49 (0) 51 32-50 08-88

Member of IVK

Company

Year of formation
1972

Size of workforce
8

Nominal capital
25.000 €

Ownership structure
GmbH & Co. KG

Sales channels
Techprint, UK

Contact partners
Management:
Carsten Branding

Application technology and sales:
Hans-Werner Klünder

Range of Products

Types of adhesives
Hot melt adhesives
Dispersion adhesives
Vegetable adhesives, dextrin and
starch adhesives
Pressure-sensitive adhesives

For applications in the field of
Paper/packaging
Bookbinding/graphic design
Wood/furniture industry
Construction industry, including floors,
walls and ceilings

BÜHNEN

Bühnen GmbH & Co. KG
Hinterm Sielhof 25
D-28277 Bremen
Phone +49 (0) 4 21-51 20-1 25
Fax +49(0) 4 21-51 20-2 60
Email: info@buehnen.de
www.buehnen.de

Member of IVK

Company

Year of formation
1922

Size of workforce
70

Ownership structure
Private ownership

Subsidiaries
PAM, Incorp., USA
Bühnen, Polska Sp. z o.o.

Contact person
Managing Director:
Hanno Pünjer

Sales Director:
Germany, Austria, Switzerland
Hans-Gerhard Hartje

Sales Director International:
Dr. Valentino Di Candido

Distribution channels
Direct sales, distributor

Range of Products

Hot melt adhesives
The product range includes a
variety of different hot melt adhesive
for almost every application.
Available bases:
EVA, PO, POR, PA, PSA, PUR, Acrylate.
Available shapes:
slugs, sticks, granules, pillows, blocks,
cartridges, barrels, drums.

Application Technology
Hot melt tank applicator systems with
piston pump or gear pump, PUR- and POR-
hot melt tank systems, PUR- and POR-bulk
unloader, hand guns for spray and bead line
application, roller coater, application heads
for beadline, surface, spray, dot, spiral
application and special application heads
with individual customer requirements,
hand-operated glue applicators, PUR- and
POR glue applicators, comprehensive range
of application accessories, customer-oriented
application, solutions.

Applications areas
Automotive, Packaging, Display Manufac-
turing, Electronic Industry, Filter Industry,
Shoe Industry, Foamplastic and Textile
Industry, Case Industry, Construction
Industry, Florists, Wood, Processing and
Furniture Industry.

Byla GmbH
Industriestraße 12
D-65594 Runkel
Phone +49 (0) 64 82-91 20-0
Fax +49 (0) 64 82-91 20-11
Email: contact@byla.de
www.byla.de

Member of IVK

Company

Year of formation
1975

Nominal capital
90.000 €

Sales channels
Worldwide

Range of Products

Types of adhesives
Reactive adhesives

For applications in the field of
Wood / furniture industry
Electronics
Mechanical engineering and equipment
construction
Automotive industry, aviation industry

Cabot Corporation

Interleuvenlaan 15i
B-3001 Leuven
Phone +32 16 39 24 00
Email: kim_herbots@cabot-corp.com
www.cabot-corp.com

Member of IVK

Company

Year of formation
1882

Size of workforce
4.000 worldwide

Managing partners
CEO: Patrick Prevost

Contact partners
Application technology and sales:
Technical Service (EMEA):
Jaume Campana,
Phone +34 93 848 43 90
Sales (EMEA):
Michael de Posch,
Phone +32 16 39 24 00

Further information
Cabot Corporation (CBT/NYSE) is a global
specialty chemical company, headquartered
in Boston, Massachusetts, USA, with sales
of $ 3.2 billion in 2008.
The company is organised in four segments:
• Core: rubber blacks and supermetals
• Performance: with specialty blacks and
  fumed metal oxides
• Service: specialty fluids
• New business: inkjet colorants, aerogel
  and superior micro powders

Range of Products

Raw materials
Additives:
Matting agents, Rheological additives,
Stabiliziers, Thixotropic agents, ...
Fillers

For applications in the field of
Wood/furniture industry
Construction industry, including floors,
walls and ceilings
Electronics
Mechanical engineering and equipment
construction
Automotive industry, aviation industry
Textile industry
Adhesive tapes, labels
Hygiene
Household, recreation and office

Celanese Emulsions GmbH
Industriepark Höchst, E416
D-65926 Frankfurt/Main
Phone +49 (0) 69-3 05-1 48 47
Fax +49 (0) 69-31 67 80
Email: Mowilith.info@celanese.de
www.celanese-emulsions.com

Member of IVK

Company

Year of formation
1863

Size of workforce
8.900 (Celanese Group)

Ownership structure
Celanese Corporation

Contact partners
Management:
Angela Cackovich (Managing Director)

Application technology and sales:
Dr. Bernhard Momper
Dr. Rolf Kuropka

Range of Products

Types of adhesives
Dispersion adhesives

Raw materials
Polymers

For applications in the field of
Paper/packaging
Bookbinding/graphic design
Wood/furniture industry
Construction industry, including floors,
walls and ceilings
Automotive industry, aviation industry
Textile industry

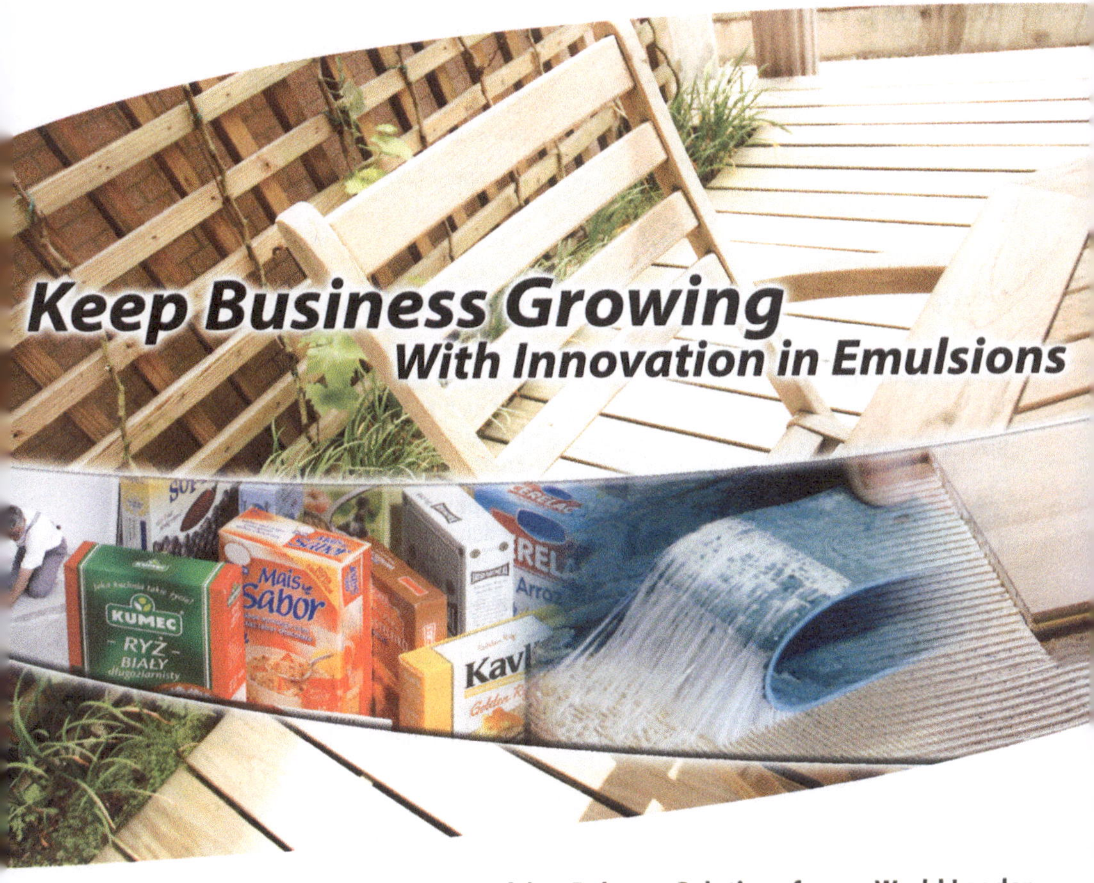

certoplast Vorwerk & Sohn GmbH

Müngstener Straße 10
D-42285 Wuppertal
Phone +49 (0) 20 2-2 55 48-0
Fax +49 (0) 20 2-2 55 48-48
Email: verkauf@certoplast.com
www.certoplast.com

Member of IVK

Company

Year of formation
1991

Size of workforce
52

Managing partners
P. Rambusch

Subsidiaries
certoplast Suzhou, China

Contact partners
Management:
P. Rambusch (General manager)

Application technology and sales:
Carsten Abt (Sales manager)

Range of Products

Types of adhesives
Hot melt adhesives
Dispersion adhesives
Pressure-sensitive adhesives

Equipment, plant and components
for adhesive curing
adhesive curing and drying

For applications in the field of
Construction industry, including floors,
walls and ceilings
Electronics
Automotive industry, aviation industry
Adhesive tapes, labels

ChemCycle Bitterfeld GmbH
ChemiePark Bitterfeld-Wolfen, Areal B
Farbenstraße 1
D-06766 Bitterfeld-Wolfen
Phone +49 (0) 34 93-7 54 00
Fax +49 (0) 34 93-7 54 04
Email:dr.r.milker@milker.de
www.chemcycle.de

Member of IVK

Company

Year of formation
2000

Size of workforce
38

Nominal capital
750.000 EUR

Ownership structure
Dr. Roland Milker 100 %

Sales channels
Direct sales to industrial partners

Contact partners
Management:
Dr. Roland Milker, CEO

Application technology and sales:
Reinhard Mainzer
Dr. Jan Marten

Further information
Toll production, Solvent Recycling,
Custom Synthesis

Range of Products

Types of adhesives
Hot melt adhesives
Solvent-based adhesives
Pressure-sensitive adhesives

Raw materials
Additives:
Crosslinker, softening agents
Polymers:
Polyacrylates solvent-based or as hot melt

For applications in the field of
Paper/packaging
Construction industry, including floors,
walls and ceilings
Textile industry
Adhesive tapes, labels

Chemetall GmbH

Trakehnerstraße 3
D-60487 Frankfurt/M.
Phone +49 (0) 69-71 65-0
Fax +49 (0) 69-71 65-29 36
www.chemetall.com

Member of IVK

Company

Year of formation
1982

Size of workforce
3.000 worldwide

Managing partners
Board of Management:
Dr. M. Engel-Bader
M. Störmer
J. Merckx

Ownership structure
A company of Rockwood Holdings Inc.

Subsidiaries
40 worldwide

Sales channels
CM subsidiaries and specific distributors

Contact partners
Management:
Thomas Willems

Application technology and sales:
Katja Gomerski
Phone +49 (0) 69-71 65-21 05

Further information
See website: www.chemetall.com
Certification to ISO 9001 : 2008,
EN 9100 : 2003, ISO 14001 : 2005

Range of Products

Types of adhesives
Hot melt adhesives
Reactive adhesives

Types of sealants
Polysulfide sealants
PUR sealants
Other epoxy

For applications in the field of
Electronics
Mechanical engineering and equipment construction
Automotive industry, aviation industry

Coim Deutschland GmbH
Novacote Flexpack Division
Schnackenburgallee 62
D-22525 Hamburg
Phone +49 (0) 40-85 31 03-0
Fax +49 (0) 40-85 31 03-69
Email: info@de.coimgroup.com
www.coimgroup.com

Member of IVK

Company

Year of formation
as COIM Group in 1962

Ownership structure
private owned company

Subsidiaries
COIM operates through a network of production sites, commercial companies and agencies located all over the world

Sales channels
The Novacote Division is part of the COIM Group, dedicated to developing and supplying adhesives and coatings, mainly for the Flexible Packaging market.

Contact partners
Management:
Volker Dreja

Application technology:
Martin Rueter

Sales:
Joerg Kiewitt

Further information
The Novacote Division is part of the Coim Group, dedicated to developing and supplying adhesives and coatings, mainly for the Flexible Packaging market. During recent years the Novacote Division grew rapidly both in terms of Business and Organization. With respect to the global organization the Novacote Technology Center is located in Hamburg, Germany as R&D Centre for Packaging.

Range of Products

Types of adhesives
Reactive adhesives
Solvent-based adhesives
Dispersion adhesives

Types of sealants
Acrylic sealants
Other

Raw materials
Resins
Polymers

For applications in the field of
Paper/packaging
Bookbinding/graphic design
Wood/furniture industry
Adhesive tapes, labels
Hygiene

Coroplast Fritz Müller GmbH & Co. KG
Wittener Straße 271
D-42279 Wuppertal
Phone +49 (0) 2 02-26 81-0
Fax +49 (0) 2 02-26 81-3 80
Email: coroplast@coroplast.de
www.coroplast.de

Member of IVK

Company

Year of formation
1928

Size of workforce
3.850

Sales channels
Wholesale and industry

Contact partners
Management:
N. Mekelburger
M. Söhngen
F. Zappe

Further information

Coroplast acts in 3 business units:
• tapes
• cables & wires
• cable assemblies

Range of Products

For applications in the field of
Paper/packaging
Wood/furniture industry
Construction industry, including floors,
walls and ceilings
Electronics
Mechanical engineering and equipment
construction
Automotive industry, aviation industry
Household, recreation and office

Cyberbond Europe GmbH
Werner-von-Siemens-Straße 2
D-31515 Wunstorf
Phone +49 (0) 50 31-95 66-0
Fax +49 (0) 50 31-95 66-26
Email: info@cyberbond.de
www.cyberbond.eu

Member of IVK

Company

Year of formation
1999

Size of workforce
15

Managing partners
Ulrich Lipper, Dieter Rademacher,
Holger Bleich, James East

Nominal capital
50.000 EUR

Ownership structure
Cyberbond LLC, USA; Ulrich Lipper,
Dieter Rademacher, Holger Bleich

Subsidiaries
Cyberbond France SARL, France
Cyberbond Iberia, Spain
Cyberbond CS s. r. o., Czech Republic

Sales channels
Direct to the industry and via exclusive
nationwide distributors as well as special
Private Label accounts

Contact partners
Management:
Ulrich Lipper, Holger Bleich

Application technology and sales:
Dieter Rademacher, Ulrich Lipper

Further information
Cyberbond –
The Power of Adhesive Information

Range of Products

Offered adhesives
Cyanoacrylates
Anaerobic Adhesives and Sealants
UV and Light Curing Adhesives
Additional programme consisting of:
Primers, Activators, D-Bonders and
Dosing Aids

Dosing equipment
LINOP Modular Dosing System for
1K Reactive Adhesives
LINOP UV LED Curing System

Products are used in
Automotive and automotive sub supplier
industry
Electronic industry
Aviation industry
Elastomer/plastic/metal working industry
Machine tool industry
Medical industry
Shoe industry
DIY, Hobby and office

DEKA (Dekalin®)
Kleben & Dichten GmbH
Gartenstraße 4
D-63691 Ranstadt
Phone +49 (0) 60 41-82 03 80
Fax +49 (0) 60 41-82 12 22

Member of IVK

Company

Year of formation
1907 DEKALIN
1999 DEKA

Size of workforce
together > 125

Ownership structure
family owned

Sales channels
industry

Contact partners
Management:
Michael Windecker

Application technology and sales:
Michael Windecker

Range of Products

Types of adhesives
Solvent-based adhesives
Dispersion adhesives

Types of sealants
Butyl sealants
PUR sealants
MS/SMP sealants

For applications in the field of
Wood/furniture industry
Construction industry, including floors,
walls and ceilings
Mechanical engineering and equipment
construction
Automotive industry, aviation industry
Household, recreation and office

DELO Industrial Adhesives

DELO-Allee 1
D-86949 Windach
Phone +49 (0) 81 93-99 00-0
Fax +49 (0) 81 93-99 00-1 44
Email: info@DELO.de
www.DELO.de

Member of IVK

Company

Year of formation
1961

Size of workforce
230

Managing partners
Dr. Wolf-Dietrich Herold
Sabine Herold

Subsidiaries
Global distributors, representative offices
in Shanghai and Singapore, subsidiary DELO
industrial adhesives LLC in the USA

Sales channels
Traders and direct

Contact partners
Application technology and sales:
Robert Saller, Director of International

Further information
DELO is a leading manufacturer of industrial
adhesives with its head office near Munich.
In the financial year 2008/09 230 employees generated a turnover of 30 million €.
The company supplies tailor-made special
adhesives and complementary equipment
for applications in special lines of business
– from electronics to the chip card and
automotive industry as well as in glass and
plastic design. DELO's customers are, for
example, Bosch, Daimler, Festo, Infineon,
NXP and Siemens. DELO has a network of
worldwide distributors and sales partners.

Range of Products

Types of adhesives
Reactive adhesives
light-curing and light-activated acrylates
and epoxies
one- and two-component epoxy resins
electrically conductive adhesives
methacrylates
cyanoacrylates
silicones
polyurethanes

Types of sealants
Acrylic sealants
PUR sealants
Silicone sealants
Other

Equipment, plant and components
for conveying, mixing, metering and for
adhesive application
for adhesive curing
measuring and testing

For applications in the field of
Construction industry, including floors,
walls and ceilings
Electronics
Mechanical engineering and equipment
construction
Automotive industry, aviation industry

DKSH GmbH

Baumwall 3
D-20459 Hamburg
Phone +49 (0) 40-37 47 34-0
Fax +49 (0) 40-37 47 34-80
Email: info@dksh.com
www.dksh.com

Company

Year of formation
1865

Size of workforce
> 20.000

Ownership structure
Privat

Contact partners
Application technology and sales:
Email: specialtychemicals.eu@dksh.com

Further information
www.dksh.com

Range of Products

Types of adhesives
Hot melt adhesives
Reactive adhesives
Solvent-based adhesives
Vegetable adhesives, dextrin and
starch adhesives
Pressure-sensitive adhesives

Raw materials
Additives:
Optical brighteners, antioxidants,
specialties
Resins:
Polyesters, PLA, specialities
Polymers:
Polyesters, PLA, specialities

For applications in the field of
Paper/packaging
Wood/furniture industry
Construction industry, including floors,
walls and ceilings
Electronics
Automotive industry, aviation industry
Textile industry
Adhesive tapes, labels

Dow Automotive

Am Kronberger Hang 4
D-65824 Schwalbach
Phone +49 (0) 61 96-5 66-0
Fax +49 (0) +49 (0) 61 96-5 66-4 16
www.dowautomotive.com

Member of IVK

Company

Ownership structure
business unit of The Dow Chemical
Company

Sales channels
direct and via distributors

Contact partners
Marc van den Biggelaar

Range of Products

Types of adhesives
1K and 2K PU and epoxy adhesives

For applications in the field of
Automotive industry, aviation industry

Dymax Europe GmbH
Trakehner Straße 3
D-60487 Frankfurt/Main
Phone +49 (0) 8 09 69-71 65-35 68
Fax +49 (0) 8 09 69-71 65-38 30
Email: dymaxinfo@dymax.de
www.dymax.com

Member of IVK

Company

Year of formation
1996

Size of workforce
17

Ownership structure
Dymax Corporation

Sales channels
Direct and Distributors

Contact partners
Management:
Dr. Detlef Heindl

Application technology and sales:
Holger Wurm
Wolfgang Lorscheider

Further information
Light-curable adhesives
Conformal coatings
Encapsulants
Maskants
Potting material
FIP/CIP-gaskets
Curing equipment

Range of Products

Types of adhesives
Reactive adhesives

Types of sealants
Acrylic sealants

Equipment, plant and components
for adhesive curing

For applications in the field of
Electronics
Automotive industry, aviation industry
Medical
Optics

Eluid Adhesive GmbH
Heinrich-Hertz-Straße 10
D-27283 Verden
Phone +49 (0) 42 31-3 03 40-0
Fax +49 (0) 42 31-3 03 40-17
Email: info@eluid.de
www.eluid.de

Member of IVK

Company

Year of formation
1932 by Emil Lumbeck
1998 by Andreas May

Size of workforce
8

Managing partners
Andreas May
Helmut May

Ownership structure
Private

Sales channels
Europe: through our own sales force,
traders and agents
Worldwide: traders and agencies

Contact partners
Management:
Andreas May

Range of Products

Types of adhesives
Hot melt adhesives
Reactive adhesives
Dispersion adhesives
Vegetable adhesives, dextrin and
starch adhesives
Glutine glue
Pressure-sensitive adhesives

Types of sealants
Acrylic sealants
Other

For applications in the field of
Paper/packaging
Bookbinding/graphic design
Adhesive tapes, labels
Hygiene
Household, recreation and office

EUKALIN Spezial-Klebstoff Fabrik GmbH

Ernst-Abbe-Straße 10
D-52249 Eschweiler
Phone +49 (0) 24 03-64 50-0
Fax +49 (0) 24 03-64 50-26
Email: eukalin@eukalin.de
www.eukalin.com

Member of IVK

Company

Year of formation
1904

Size of workforce
49 employees

Managing partners
Dr. Joachim Schulz
Jürgen Schulz-Wachler
Timm Koepchen

Ownership structure
family owned

Subsidiaries
EUKALIN Corp. USA

Sales channels
Directly and through dealers

Contact partners
Application technology and sales:
Timm Koepchen

Range of Products

Types of adhesives
Hot melt adhesives
Dispersion adhesives
Vegetable adhesives, dextrin and
starch adhesives
Glutine glue
Pressure-sensitive adhesives

Equipment, plant and components
for conveying, mixing, metering and
for adhesive application

For applications in the field of
Paper/packaging
Bookbinding/graphic design
Textile industry

Evonik Industries
Coatings & Additives
D-45764 Marl, www.coatings-colorants.com
www.evonik.com/crosslinkers,
www.evonik.com/adhesives
Inorganic Materials
D-60287 Frankfurt/M., www.aerosil.com,
www.dynasylan.com, www.evonik.com/fp
Performance Polymers
D-45764 Marl, www.vestamelt.de
Consumer Specialities
D-45127 Essen, www.evonik.com/polymer-dispersions
Member of IVK

Company

Year of formation
2001

Ownership structure
74.9 % RAG Stiftung, 25.1 % CVC

Subsidiaries
Evonik Degussa GmbH, Evonik Goldschmidt GmbH

Contact partners
Application technology and sales:
Inorganic Materials:
Phone +49 (0) 76 23-91-83 92
(application techn.)
Phone +49 (0) 69-2 18-34 76 (sales)
Email: aerosil@evonik.com, fillers.pigments@
evonik.com
Coatings & Additives:
Phone +49 (0) 23 65-49-48 43
Fax +49 (0) 23 65-49-50 30
Email: co@evonik.com
Performance Polymers:
Phone +49 (0) 23 65-49-43 56
Fax +49 (0) 23 65-49-59 92
Email: vestamelt@evonik.com
Consumer Specialties:
Phone +49 (0) 2 01-1 73 27 06
(application techn.)
Fax +49 (0) 2 01-1 73 29 62 (sales)
Email: info@polymerdispersion.com

Further information
Evonik Industries operates in three business areas:
Chemicals, Energy and Real Estate. Evonik is a
global leader in specialty chemicals, an expert in
power generation from hard coal and renewable
energies, and one of the largest private residential
real estate companies in Germany. Evonik is active
in over 100 countries around the world. In its fiscal
year 2008 about 41,000 employees generated sales
of about € 15.9 billion and an operating profit
EBITDA of about € 2.2 billion.

Range of Products

Types of adhesives
Hot melt adhesives (VESTAMELT®) (DYNAPOL® S)

Types of sealants
Acrylic sealants

Raw materials
Additives: Waxes (VESTOWAX®),
Defoamer (TEGO® Antifoam),
Wetting agents (TEGOPREN®, TEGO® Surten),
Thickener (TEGO® Rheo),
Dispersant (TEGOMER® DA),
functional siloxanes for reactive adhesives

Fillers: pyrogenic silicas and metal oxides
(AEROSIL®), specialty precipitated silicas
(SIPERNAT®), pigment blacks (PRINTEX®),
functional silanes (Dynasylan®)

Resins: speciality resins, aliphatic diamines
(VESTAMIN®), aliphatic isocyanates (VESTANAT®)

Solvents

Polymers: amorphous poly-alpha-olefines
(VESTOPLAST®); copolyesters (DYNAPOL®,
DYNACOLL®); liquid polybutadienes
(Polyoil, POLYVEST®); polyacrylates (DEGALAN®)

For applications in the field of
Paper/packaging
Bookbinding/graphic design
Wood/furniture industry
Construction industry, including floors,
walls and ceilings
Electronics
Automotive industry, aviation industry
Textile industry
Adhesive tapes, labels
Hygiene
Manufacturing of hotmelt adhesives

ExxonMobil
Chemical Central Europe GmbH
Neusser Landstraße 16
D-50735 Köln
Phone +49 (0) 2 21-77 03-0
www.exxonmobil.de

Member of IVK

Company

Year of formation
1999 – Fusion Exxon and Mobil zur
ExxonMobil Corporation

Contact partners
Management Germany:
Wolfgang Rössing
Phone +49 (0) 2 21-77 03-2 36
Email: wolfgang.roessing@exxonmobil.com
Reimond Rohde
Phone +49 (0) 2 21-77 03-2 34
Email: reimond.rohde@exxonmobil.com

Application technology and sales:
see above

Range of Products

Types of adhesives
Hot melt adhesives
Solvent-based adhesives

Types of sealants
Butyl sealants
Other

Raw materials
Resins
Solvents
Polymers

For applications in the field of
Paper/packaging
Bookbinding/graphic design
Wood/furniture industry
Construction industry, including floors,
walls and ceilings
Electronics
Mechanical engineering and equipment
construction
Automotive industry, aviation industry
Textile industry
Adhesive tapes, labels
Hygiene
Household, recreation and office

fischer Deutschland Vertriebs GmbH

Weinhalde 14 – 18
D-72178 Waldachtal
Phone +49 (0) 74 43 12-0
Fax +49 (0) 74 43 12-42 22
www.fischer.de

Member of IVK

Company

Year of formation
1948

Size of workforce
3.800

Ownership structure
Privately owned

Subsidiaries
30 (Argentina, Australia, Blegium, Brasil,
China, Denmark, Germany, Finland, France,
Greece, Great Britain, Italy, Japan, Croatia,
Mexico, South Korea, Czech Republic,
Hungary, USA, United Arab Emirates

Sales channels
DIY, specialized trade

Contact partners
Management:
Thomas Lopian
Email: thomas.Lopian@fischer.de

Further information
World market leader in chemical fastening
systems

Range of Products

Types of adhesives
Reactive adhesives
Solvent-based adhesives
Dispersion adhesives

Types of sealants
Acrylic sealants
Silicone sealants
MS/SMP sealants

For applications in the field of
Wood/furniture industry
Construction industry, including floors,
walls and ceilings
Household, recreation and office

Follmann & Co. GmbH & Co. KG

Karlstraße 59
D-32423 Minden
Phone +49 (0) 5 71-93 39-0
Fax +49 (0) 5 71-93 39-3 00
Email: info@follmann.de
www.follmann.de

Member of IVK

Company

Year of formation
1977

Size of workforce
276 (2008)

Managing partners
Stefan Wernecke
Dr. Thomas Damerau

Subsidiaries
OOO Follmann
(Egorewsk/Russian Federation)

Contact partners
Management:
Thomas Bierwirth

Application technology and sales:
Torsten Krite

Range of Products

Types of adhesives
Hot melt adhesives
Dispersion adhesives

Raw materials
Polymers

For applications in the field of
Paper/packaging
Bookbinding/graphic design
Wood/furniture industry
Construction industry, including floors,
walls and ceilings

BONDING SYSTEMS

Forbo Adhesives Deutschland GmbH
Zweibruecker Straße 185
D-66954 Pirmasens
Phone +49 (0) 63 31-5 28-0
Fax +49 (0) 63 31-5 28-2 01
Email: info-klebstoffe@forbo.com
www.forbo.com

Member of IVK

Company

Year of formation
1919

Size of workforce
150

Contact partners
Patricio Lacalle, General Manager

Range of Products

Types of adhesives
Hot melt adhesives
Reactive adhesives
Solvent-based adhesives
Dispersion adhesives
Vegetable adhesives, dextrin and
starch adhesives
Pressure-sensitive adhesives

Types of sealants
PUR sealants
MS/SMP sealants

For applications in the field of
Paper/packaging
Bookbinding/graphic design
Wood/furniture industry
Construction industry, including floors,
walls and ceilings
Automotive industry, aviation industry
Textile industry

Forbo Erfurt GmbH

August-Röbling-Straße 2
D-99091 Erfurt
Phone +49 (0) 3 61-7 30 41-0
Fax +49 (0) 3 61-7 30 41-91
Email: info.erfurt@forbo.com
www.erfurt.forbo.com

Member of IVK

Company

Year of formation
1919/1920

Size of workforce
90

Managing partners
Forbo Beteiligungen GmbH,
D-79539 Lörrach

Nominal capital
2.050.000 EUR

Ownership structure
100 % share holder

Sales channels
direct sales and/or through agents

Contact partners
Management:
Ruediger Beez

Application technology and sales:
Ruediger Beez

Range of Products

Types of adhesives
Reactive adhesives
Solvent-based adhesives
Dispersion adhesives
Pressure-sensitive adhesives

For applications in the field of
Construction industry, including floors,
walls and ceilings

Please also mark your calendar:
FEICA European Adhesives &
Sealants Conference 2010
September 16–17 · Scandic Marina Center
Helsinki · Finland

FEICA European Adhesives & Sealants Conference 2009

September 17-18 · Hotel InterContinental · Budapest · Hungary

The place to be!

www.feica-conferences.com

FEICA
Association of European Adhesives & Sealants Manufacturers
Conference Management: Marion Krämer, T: +49-2 11-69 55 60 04
E: marion.kraemer@feica-conferences.com

**H. B. Fuller
Austria GesmbH**
Kaplanstraße 30
A-4600 Wels
Email: info-europe@hbfuller.com
www.hbfuller.com

Member of IVK

Company

Year of formation
1887

Size of workforce
3.200

Managing partners
James McCreary, Cheryl Reinitz, Ralf Müller, Stefan Reuss, Hilpert Kostka

Other subsidiaries
H. B. Fuller France S. A., H. B. Fuller Italia Srl., H. B. Fuller Benelux B. V., H. B. Fuller Espana S. A., H. B. Fuller Isar Rakoll SA, H. B. Fuller UK Ltd., H. B. Fuller Sverige AB, H. B. Fuller Egypt

Sales channels
Direct sales, Distribution companies

Further information
H. B. Fuller is leading manufacturer that provides specialtity chemical products and services to a variety of markets, including construction, packaging and personal care. Our adhesives, sealants and other products are used to create hundreds of brand name goods across the globe.
The company's corporate headquarters and a 225.000 square-foot R & D facility are located in St. Paul, Minn. In Europe we have plants, service centres and regional product development labs in Austria, Germany, the UK, Italy and Portugal.

Range of Products

Types of adhesives
Hot melt adhesives
Reactive adhesives
Solvent-based adhesives
Dispersion adhesives
Vegetable adhesives, dextrin and starch adhesives
Pressure-sensitive adhesives

Types of sealants
Butyl sealants
Polysulfide sealants
PUR sealants

For applications in the field of
Paper/packaging
Bookbinding/graphic design
Wood/furniture industry
Construction industry, including floors, walls and ceilings
Electronics
Mechanical engineering and equipment construction
Automotive industry, aviation industry
Textile industry
Adhesive tapes, labels
Hygiene

**H. B. Fuller
Deutschland GmbH**
An der Roten Bleiche 2 - 4,
D-21335 Lüneburg
Henriettenstraße 32,
D-31582 Nienburg/W.
Email: info-europe@hbfuller.com
www.hbfuller.com

Member of IVK

Company

Year of formation
1887

Size of workforce
3.200

Managing partners
James McCreary, Cheryl Reinitz, Ralf Müller,
Stefan Reuss, Hilpert Kostka

Other subsidiaries
H. B. Fuller France S. A., H. B. Fuller Italia Srl.,
H. B. Fuller Benelux B. V., H. B. Fuller Espana
S. A., H. B. Fuller Isar Rakoll SA, H. B. Fuller UK
Ltd., H. B. Fuller Sverige AB, H. B. Fuller Egypt

Sales channels
Direct sales, Distribution companies

Further information
H. B. Fuller is leading manufacturer that
provides specialtity chemical products and
services to a variety of markets, including
construction, packaging and personal care.
Our adhesives, sealants and other products
are used to create hundreds of brand name
goods across the globe.
The company's corporate headquarters and
a 225.000 square-foot R & D facility are
located in St. Paul, Minn. In Europe we have
plants, service centres and regional product
development labs in Austria, Germany, the
UK, Italy and Portugal.

Range of Products

Types of adhesives
Hot melt adhesives
Reactive adhesives
Solvent-based adhesives
Dispersion adhesives
Vegetable adhesives, dextrin and
starch adhesives
Pressure-sensitive adhesives

Types of sealants
Butyl sealants
Polysulfide sealants
PUR sealants

For applications in the field of
Paper/packaging
Bookbinding/graphic design
Wood/furniture industry
Construction industry, including floors,
walls and ceilings
Electronics
Mechanical engineering and equipment
construction
Automotive industry, aviation industry
Textile industry
Adhesive tapes, labels
Hygiene

GLUDAN Deutschland GmbH

Auf der Geest 6
D-21514 Büchen
Phone +49 (0) 41 55-20 27
Fax +49 (0) 41 55-38 41
Email: gludan@gludan.de
www.gludan.com

Member of IVK

Company

Year of formation
1977

Size of workforce
20

Ownership structure
GmbH

Sales channels
Own sale force, and agents, have a look on
our map www.gludan.com for contacts

Contact partners
Management:
is@gludan.de
ks@gludan.dk

Application technology and sales:
am@gludan.de
jr@gludan.de
sale-ow@gludan.dk
sb@gludan.de

Further information
www.gludan.com

Range of Products

Types of adhesives
Hot melt adhesives
Dispersion adhesives
Glutine glue
Pressure-sensitive adhesives

Raw materials
Additives
Fillers
Polymers
Starch

Equipment, plant and components
for conveying, mixing, metering and
for adhesive application
measuring and testing

For applications in the field of
Paper/packaging
Bookbinding/graphic design
Wood/furniture industry
Construction industry, including floors,
walls and ceilings
Textile industry
Adhesive tapes, labels
Hygiene
Household, recreation and office

hanse chemie AG
Charlottenburger Straße 9
D-21502 Geesthacht
Phone +49 (0) 41 52-80 92-0
Fax +49 (0) 41 52-7 91 56
Email: info@hanse-chemie.com
www.hanse-chemie.com

Member of IVK

Company

Looking to meet customer needs and maximize the benefit of our extensive R & D efforts, we respond to each enquiry in a direct dialogue with the customer to jointly assess the requirements of the intended application. Based on this input, specific solutions are developed and refined to create trail-blazing products. Enhanced by our association with renowned universities, research institutes and international partners, we develop innovative premium base materials and specialties for adhesives, sealants as well as silicone formulations. The demand for silicone products is steadily increasing worldwide – e. g. in the construction, automotive and shipbuilding industries and the dental sector.

Range of Products

Raw materials
Additives:
Silane crosslinkers, modifiers, catalysts
Polymers:
PUR, silicones, silane terminated polymers (STP)

For applications in the field of
Construction industry, including floors, walls and ceilings
Electronics
Automotive industry
Aviation industry

**Henkel
AG & Co. KGaA**
Henkelstraße 67
D-40191 Düsseldorf
Phone +49 (0) 2 11-7 97-0
www.henkel.com

Member of IVK

Company

Ownership structure
AG & Co. KGaA

Contact partners
Management:
Business Unit Adhesive Technologies
Phone +49 (0) 2 11-7 97-61 91
Fax +49 (0) 2 11-7 98-1 61 91

Further information
Henkel is the world market leader in
adhesives, sealants and surface treatments
for consumers, craftsmen and industrial
applications.
From our long tradition we have a well-
balanced portfolio of international, regional
and local brands. Henkel offers a multitude
of applications to satisfy the needs of
different target groups – consumers as well
as craftsmen and industrial business.
The Adhesive Technologies business sector
serves a wide range of user groups with a
comprehensive range of technologies:
Adhesive and sealant systems for industrial
applications in the automotive, packaging,
aircraft, electronics, durable consumer
goods and metal sectors, and for mainte-
nance, repair and overhaul applications.
Building adhesives for architects and
building contractors, adhesives and sealants
for craftsmen, DIY users and consumers
for applications in the home, school and
office. Our Top Brands are Loctite, Teroson,

Range of Products

Types of adhesives
Hot melt adhesives
Reactive adhesives
Solvent-based adhesives
Dispersion adhesives
Vegetable adhesives, dextrin and
starch adhesives
Pressure-sensitive adhesives

Types of sealants
Acrylic sealants
Butyl sealants
Polysulfide sealants
PUR sealants
Silicone sealants
MS/SMP sealants
Other

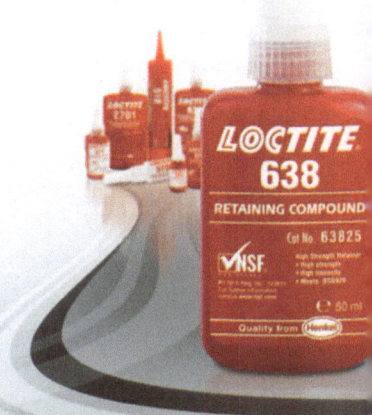

Hexion Specialty Chemicals B.V.
Koddeweg 67
NL-3194 DH Hoogvliet/Rotterdam
Phone +31 (0) 295 41 00
Email: jurgen.vanderlinden@hexion.com
www.hexion.com

Member of IVK

Company

Year of formation
2005

Ownership structure
Apollo Management LP

Sales channels
direct to end-users

Contact partners :
Francisco Reig
BU Manager Dispersions for
Coating & Adhesives

Application technology and sales:
Jurgen van der Linden
Marketing Manager
Phone +31 (0) 6-53 67 42 15

Further information
Founded after acquisition by
Apollo Inv. Company

Range of Products

Acrylic and Styrene Acrylic based polymeric
dispersions for Coatings, Adhesives and
Building & Construction

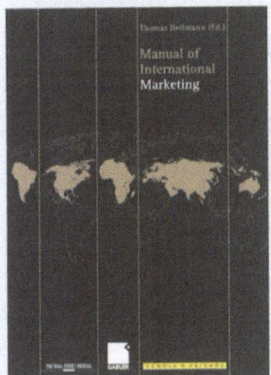

Enriching lives through innovation

Huntsman Advanced Materials (Switzerland) GmbH
Klybeckstraße 200, CH-4057 Basel
Phone +41 61-2 99 11 11
Fax +41 61-2 99 11 12
Email: advanced_materials@huntsman.com
www.huntsman.com/advanced_materials
www.araldite.com

Member of IVK

Company

Huntsman Advanced Materials is a division of Huntsman, a global manufacturer and marketer of differentiated chemicals. Huntsman today has more than 12,000 employees and operates from multiple locations worldwide. The company had 2008 revenues exceeding US $ 10 billion.

Huntsman Advanced Materials is a pioneer and marketer of advanced epoxy resins, adhesives, electrical insulation, electronics, tooling and modeling materials, and structural composites.

Facts & Figures Huntsman Advanced Materials:
• US $ 1,438 Mio Revenue in 2008
• 2,200 empolyees
• global sales in more than 90 countries
• over 4,500 customers

Contacts

Europe/Middle East/Africa
Dr. Joachim Sauer, Marketing Manager Adhesives
Phone +41 61-2 99 25 57
Email: joachim_sauer@huntsman.com

APAC/India
Chun Tak Hui, Sales Development Manager
Phone +852 2148 8892
Email: chun_tak_hui@huntsman.com

Americas
Doug Ellerbusch, Marketing Manager Adhesives
Phone +1 281 719 6060
Email: doug_ellerbusch@huntsman.com

Regional headquarters

Europe/Middle East/Africa
Huntsman Advanced Materials (Switzerland) GmbH
Klybeckstraße 200, CH-4057 Basel
Phone +41 61-2 99 11 11, Fax +41 61-2 99 11 12

Range of Products

Huntsman Advanced Materials is a global leader in the development and supply of Araldite® high performance adhesives. Araldite® adhesives cover a broad range of applications and offer both production and performance benefits to industrial, consumer and construction markets.

Types of adhesives
Epoxy adhesives
Polyurethane adhesives
Methacrylate adhesives
UV-curing adhesives
Hotmelts

For applications in the field of
Aerospace, Automotive, Construction, Consumer, Electronics, Engineering, Filters, General Industry, Household, Marine, Medical Engineering, Metal Processing, Plastic Processing, Power Transmission & Distribution, Solar Energy, Sports & Leisure, Wind Energy

Sales
Direct Sales, Distribution, Specialist Shops

APAC/India
Huntsman Advanced Materials (Hong Kong) ltd.
Suites 3 – 12, Level 41, Langham Place
8 Argyle Street, Kowloon, Hong Kong
Phone +852 2148 8800, Fax +852 2424 1741

Americas
Huntsman Advanced Materials (Americas) Inc.
10003 Woodloch, Forest Drive
The Woodlands, Texas 77380
Phone + 1 888 564 9318, Fax +1 281 719 4047

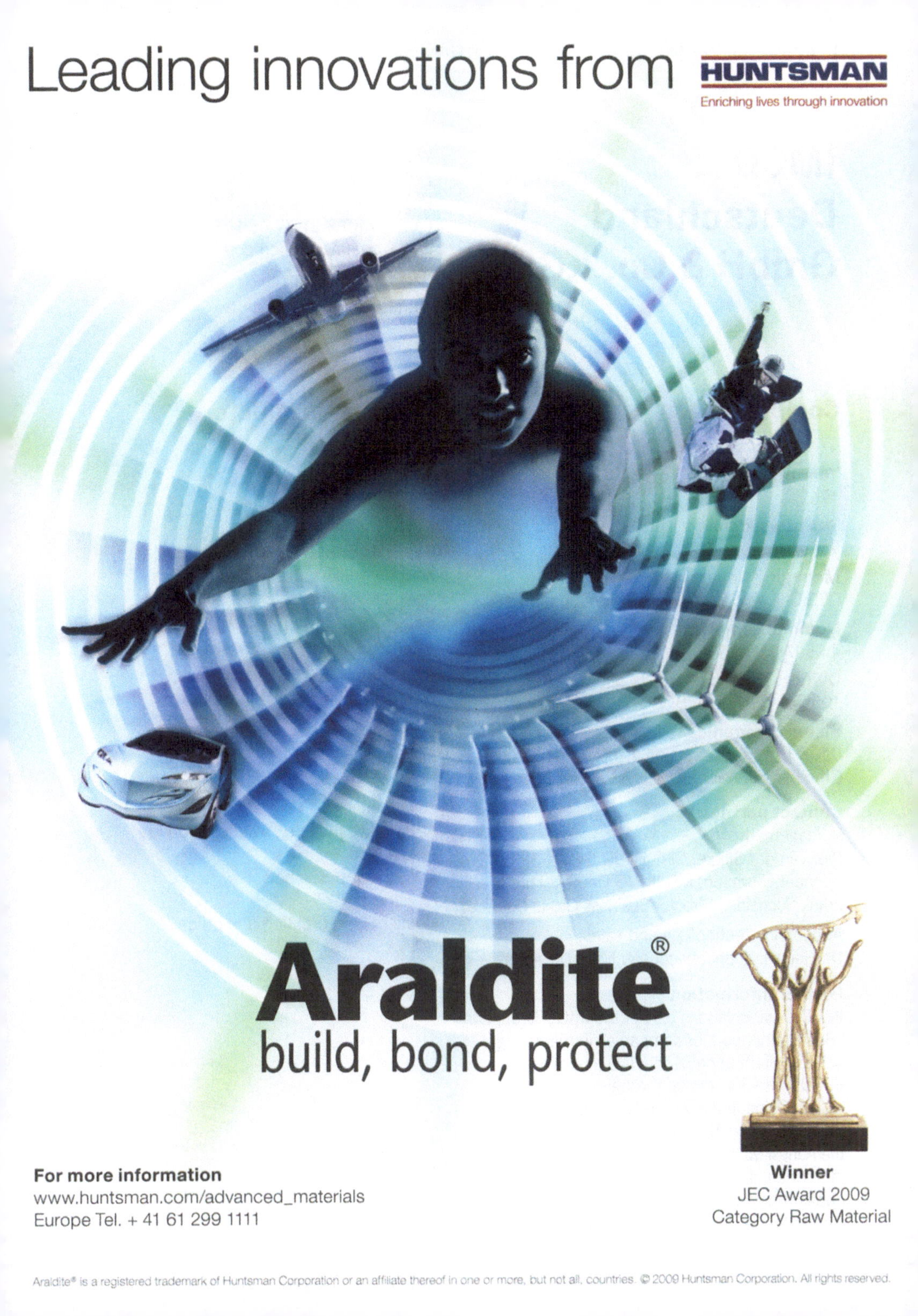

Leading innovations from

HUNTSMAN
Enriching lives through innovation

Araldite®
build, bond, protect

Winner
JEC Award 2009
Category Raw Material

For more information
www.huntsman.com/advanced_materials
Europe Tel. + 41 61 299 1111

IMCD Deutschland GmbH & Co. KG

Woerthstraße 34
D-50668 Köln
Phone +49 (0) 2 21-77 65-0
Fax +49 (0) 2 21-77 65-3 05
Email: coatings@imcd.com
www.imcdgroup.com

Member of IVK

Company

Year of formation
1960

Size of workforce
95

Managing partners
CBG Chemie Beteiligungsgesellschaft mbH, Köln

Nominal capital
4,09 Mill. €

Ownership structure
100 % IMCD Group B. V.

Sister Companies
IMCD Group operates all over Europe, Africa, China, India, Australia & New Zealand

Contact partners
Management:
Piet van der Slikke, CEO IMCD Group
Michael Thomson, MD IMCD Deutschland
Frank Schneider, Director Business Group Coatings

Application technology and sales:
Dr. Heinz-J. Küppers, Industry Manager Adhesives

Further information
IMCD Deutschland is part of the IMCD Group, a leading network of companies in marketing, sales & distribution of specialty chemicals. IMCD offers its service throughout Europe, Africa, China, India, Australia & New Zealand. IMCD Coatings Business Group is dedicated to the Adhesives, Inks, Coatings, Paints & Construction industries. IMCD Coatings operates three application labs and has set up a Technical Competence Center to support customers and suppliers.

Range of Products

Types of adhesives
Solvent-based adhesives,
Dispersion adhesives

Raw materials
Additives, Fillers, Resins, Solvents, Polymers, Starch

For applications in the field of
Paper/packaging
Bookbinding/graphic design
Wood/furniture industry
Construction industry, including floors, walls and ceilings
Electronics
Mechanical engineering and equipment construction
Automotive industry, aviation industry
Textile industry
Adhesive tapes, labels
Hygiene
Household, recreation and office

Jowat AG

Ernst-Hilker-Straße 10 – 14
D-32758 Detmold
Phone +49 (0) 52 31-7 49-0
Fax +49 (0) 52 31-7 49-1 05
Email: info@jowat.de
www.jowat.de

Member of IVK

Company

Year of formation
1919

Size of workforce
approx. 650

Managing partners
Klaus Kullmann
Ralf Nitschke
Dr. Christian Terfloth

Ownership structure
Shareholder company (not publicly traded)

Subsidiaries
17 worldwide

Sales channels
Own affiliations and distributors

Contact partners
Management:
Division Wood, furniture, construction:
Ulrich Schaefer

Division Paper, Packaging, Graphics Industry:
Manfred Kubo

Division Automotive, Textile:
Dr. Thomas Hohberg

Range of Products

Types of adhesives
Hot melt adhesives
Reactive adhesives
Solvent-based adhesives
Dispersion adhesives
Pressure-sensitive adhesives

For applications in the field of
Paper/packaging
Bookbinding/graphic design
Wood/furniture industry
Construction industry, including floors, walls and ceilings
Electronics
Automotive industry, aviation industry
Textile industry
Adhesive tapes, labels

A score of components to bond.
A score of adhesives to bond them.
Make sure you score: bond with Jowat.

Tomorrow's adhesives today.

Jowat Klebstoffe GmbH

Dr.-Pier-Straße 1
D-06729 Elsteraue (OT Göbitz)
Phone +49 (0) 34 41-53 82-0
Fax +49 (0) 34 41-53 82-2 50
Email: info@jowat.de
www.jowat.de

Member of IVK

Company

Year of formation
1992

Size of workforce
approx 70

Managing partners
Jowat AG

Sales channels
Affiliations and distributions

Contact partners
Managing Director:
Ralf Nitschke

Plant Manager:
Uwe Lehmeier

Business Manager:
Gabriele Müller

Range of Products

Types of adhesives
Hot melt adhesives
Solvent-based adhesives
Dispersion adhesives
Pressure-sensitive adhesives
Separating agents

For applications in the field of
Paper/packaging
Bookbinding/graphic design
Wood/furniture industry
Construction industry, including floors,
walls and ceilings
Automotive industry, aviation industry
Upholstery, mattresses

Five continents –
one source of knowledge

Sean H., Manufacturing Manager
AUSTRALIA

Daniel G., Application Engineer
EUROPE

Vanessa P., Executive Director Product Engineering
AMERICA

Antony F., Director of Business Development
AFRICA

Ming Li H., Senior Electronics Engineer
ASIA

Drawing on the credibility and longevity of ATZ, the longest established auto trade magazine in the world, AutoTechnology has been rebranded ATZautotechnology.

ATZautotechnology delivers great value. With increased content, more issues per year and greater geographical coverage, including China and India, it is still the best source of technical information for global automotive engineering, production and management teams.

For more information, visit our website www.ATZonline.com

Magazine of FISITA – the world body for automotive engineers

Kaneka Belgium N.V.
Liquid Polymers Division
Boulevard du Triomphe 173
B-1160 Brussels
Phone +32 2-6 63 01 95
Fax +32 2-6 63 01 84
Email: info.LPD@Kaneka.be
www.kaneka.be

Member of IVK

Company

Year of formation
Kaneka Belgium N.V. was founded in 1970 as the European production site of the globally acting Kaneka Corporation, Japan.

Further information
Kaneka products have conquered the European market, becoming a synonym for premium quality raw materials, with the brand MS POLYMER basically defining a new group of adhesives and sealants. Other brands of Kaneka's Liquid Polymer Division are SILYL, EPION, and XMAP.

Production/Technical Service contact
Kaneka Belgium N.V.
Liquid Polymers Division
Nijverheidsstraat 16
B-2260 Westerlo-Oevel
Phone +32 14-25 78 00
Fax +32 14-25 78 87

Marketing contact
Kaneka Belgium N.V.
Liquid Polymers Division
Boulevard du Triomphe 173
B-1160 Brussels
Phone +32 2-6 63 01 95
Fax +32 2-6 63 01 84
Email: info.LPD@Kaneka.be

For contact in D, A, CH, CEE
Kunststoffe Werner Hollbeck GmbH
Kirchmannstraße 22
D-45133 Essen
Phone +49 (0) 2 01-7 22 16 17
Email: info@hollbeck.de

Range of Products

Raw materials
MS Polymer, SILYL, EPION, and XMAP are reactive liquid polymers based on polyether, polyisobutene, or polyacrylate, respectively. Customers can select between moisture curing grades, addition cure types, and radical cure types including UV-cure. The polymers' characteristics produce elastic adhesives and sealants, but also high strength types and coatings can be formulated.

End products' features
Solvent and isocyanate free 1-p/2-p
Pressure sensitive adhesives (PSA)
Oil resistance, temperature resistance
(150 °C permanent use)
Low gas and moisture permeability
High UV-resistance
Adhesive blends with epoxies

For applications in the field of
Wood/furniture industry
Construction industry, including floors, walls and ceilings
Electronics
Automotive industry, transportation industry
Adhesive tapes, labels
Household, recreation and office
General industry
Shipbuilding industry

Kiesel Bauchemie GmbH & Co. KG

Wolf-Hirth-Straße 2
D-73730 Esslingen
Phone +49 (0) 7 11-9 31 34-0
Fax +49 (0) 7 11-9 31 34-1 40
Email: kiesel@kiesel.com
www.kiesel.com

Member of IVK

Company

Year of formation
1959

Size of workforce
130

Ownership structure
family owned

Subsidiaries
Sales offices in The Netherlands, Poland,
Czech Republic

Sales channels
wholesale to professional installers

Contact partners
Management:
Beatrice Kiesel

Application technology and sales:
Ulrich Lauser

Range of Products

Types of adhesives
Reactive adhesives
Solvent-based adhesives
Dispersion adhesives

Types of sealants
Acrylic sealants
PUR sealants

For applications in the field of
Construction industry, including floors,
walls and ceilings

Kisling AG

Ringstraße 30
CH-8317 Tagelswangen
Phone +41 (0) 52 354 33 33
Fax +41 (0) 52 354 33 03
Email: info@kisling.com
www.kisling.com

Member of IVK

Year of formation
1862

Size of workforce
52

Managing partners
4

Nominal capital
1.500.000 CHF

Contact partners
Management:
Hanspeter Tanner, CEO

Application technology and sales:
Gianpietro Belotti, Sales and Marketing
Manager

Further information
www.kisling.com

Types of adhesives
Reactive adhesives

Types of sealants
Acrylic sealants
Butyl sealants
Silicone sealants
MS/SMP sealants
Other

Equipment, plant and components
for conveying, mixing, metering and
for adhesive application

For applications in the field of
Wood/furniture industry
Construction industry, including floors,
walls and ceilings
Electronics
Mechanical engineering and equipment
construction
Automotive industry, aviation industry
Textile industry
Hygiene
Household, recreation and offic

Klebstoffwerke Collodin GmbH & Co. KG

Vilbeler Landstraße 20
D-60386 Frankfurt/M.
Phone +49 (0) 69-4 01 04-0
Fax +49 (0) 69-4 01 04-1 15
Email: info@collodin.de
www.collodin.de

Member of IVK

Company

Year of formation
1875

Sales channels
direct and agencies

Contact partners
Management:
Paul Joseph Nauth
Dr. Manfred Dietz

Application technology and sales:
Dr. Manfred Dietz

Range of Products

Types of adhesives
Hot melt adhesives
Reactive adhesives
Dispersion adhesives
Casein, dextrin and starch adhesives
Pressure-sensitive adhesives

Raw materials
Rosins: special rosin soaps and dispersions

Equipment, plant and components
for mixing, metering and chemical reactions
for production of adhesives and special
chemical components
for measuring and testing

For applications in the field of
Paper converting/packaging/labelling
Bookbinding/graphic design
Wood/furniture industry
Electronics
Mechanical engineering and equipment
construction
Automotive industry
Adhesive tapes, labels

**Klebchemie M. G. Becker
GmbH & Co. KG**
Max-Becker-Straße 4
D-76356 Weingarten
Phone +49 (0) 7 24 46 20
Fax +49 (0) 7 24 47 00-0
Email: info@kleiberit.com
www.kleiberit.com

Member of IVK

Company

Year of formation
1948

Size of workforce
300

Ownership structure
Owner Managed GmbH & Co. KG

Subsidiaries
Australia, France, USA, Canada, UK, Japan,
China, Singapore, Russia

Sales channels
Direct, Wholesale

Contact partners
Management:
Dipl. Phys. Klaus Becker-Weimann

Application technology and sales:
Elke Ehrmann

Further information
Specialist in PUR-Adhesive-Technology

Range of Products

Types of adhesives
Hot melt adhesives
Reactive adhesives
Solvent-based adhesives
Dispersion adhesives
Pressure-sensitive adhesives

Types of sealants
Acrylic sealants
PUR sealants
Silicone sealants
MS/SMP sealants

For applications in the field of
Wood/furniture industry
Construction industry, including floors,
walls and ceilings
Mechanical engineering and equipment
construction
Automotive industry, aviation industry
Textile industry

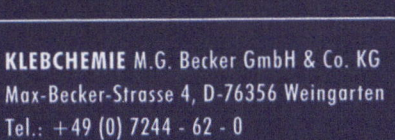

KÖMMERLING CHEMISCHE FABRIK GMBH

Kömmerling Chemische Fabrik GmbH
Zweibrücker Straße 200
D-66954 Pirmasens
Phone +49 (0) 63 31 56-20 00
Fax +49 (0) 63 31 56-19 99
Email: klebstoff@koe-chemie.de
www.koe-chemie.de

Company

Year of formation
1897

Size of workforce
400

Ownership structure
Subsidiary of ADCO Europe Holding GmbH

Subsidiaries
Kömmerling Chimie SARL, Strasbourg (F)
Kommerling UK LTD, Uxbridge (UK)

Sales channels
Industry, trading

Contact partners
Management:
José C. Houssa
Charles R. Spalton

Application technology and sales:
Sales industrial adhesives:
Dr. Gert Heckmann;
Sales Glass:
Herbert Haas
Application Technology:
Dr. Norbert Schott

Further information
Kömmerling Chemische Farbrik GmbH is a
leading international manufacturer of high
quality adhesives and sealants. Established
over 110 years ago, Kömmerling is today a
major systems supplier for the automotive,
insulating glass, window, construction,
footwear, furniture, photovoltaic and solar
thermal industry.

Range of Products

Types of adhesives
Hot melt adhesives
Reactive adhesives
Solvent-based adhesives
Dispersion adhesives
Pressure-sensitive adhesives

Types of sealants
Butyl sealants
Polysulfide sealants
PUR sealants
Silicone sealants
MS / SMP sealants

For applications in the field of
Construction industry, including floors,
walls and ceilings
Automotive industry, aviation industry
Adhesive tapes, labels
Marine, coil coating, shoe industry, insulat-
ing glass, photovoltaic, solar-thermal, direct
glazing, structural glazing

Krahn Chemie GmbH
Grimm 10
D-20457 Hamburg
Phone +49 (0) 40-3 20 92-0
Fax +49 (0) 40-3 20 92-3 22
Email: info@krahn.de
www.krahn.de

Member of IVK

Company

Year of formation
1972

Size of workforce
75

Managing directors
Dr. Jörg Schottek
Axel Sebbesse

Nominal capital
3.07 Mio €

Ownership structure
Otto Krahn GmbH & Co. KG founded 1909

Subsidiaries
Albis Plastic GmbH is a subsidiary of
Otto Krahn GmbH & Co. KG

Contact partners
Management:
Bernd Sparenberg
Email: b.sparenberg@krahn.de

Application technology and sales:
Sven Thomas
Email: s.thomas@krahn.de

Range of Products

Raw materials
Additives & Auxiliaries
Fillers
Resins: Hydrocarbonic, Epoxy, Phenolic
Modifiers
Anhydrides
Solvents
Polymers, Dispersions & Emulsions
Waxes
CR-Latices
CR/CSM
Metal soaps
Plasticizers
Pigment & Anticorrosive Pigments
EVA
PU
Adhesion Promoters

For applications in the field of
Paper/packaging
Bookbinding/graphic design
Wood/furniture industry
Construction industry, including floors,
walls and ceilings
Electronics
Automotive industry, aviation industry
Textile industry
Adhesive tapes, labels

Kraton Polymers GmbH

Mergenthalerallee 15 – 21
D-65760 Eschborn
Phone +49 (0) 61 96-7 77 17-0
Fax +49 (0) 61 96-7 77 17-11
www.kraton.com

Member of IVK

Company

Contact partners
Management
Customer.Inquiries@Kraton.com

Application technology and sales
Customer.Inquires@Kraton.com

Further information
Kraton is a leading global producer of
engineered polymers and styrenic block
copolymers (SBC's), a familiy of products
whose chemistry was pioneered by us over
forty years ago.

Range of Products

Types of adhesives
Hot melt adhesives
Solvent-based adhesives
Pressure-sensitive adhesives

Raw materials
Polymers: Styrenic block copolymers,
polyisoprene

For applications in the field of
Paper/packaging
Bookbinding/graphic design
Wood/furniture industry
Construction industry, including floors,
walls and ceilings
Automotive industry, aviation industry
Textile industry
Adhesive tapes, labels
Hygiene

LANXESS Deutschland GmbH
Kaiser-Wilhelm-Allee
D-51369 Leverkusen
Phone +49 (0) 2 14-3 04 91 89
Email: michael.herrmann@lanxess.com
www.lanxess.com

Member of IVK

Company

Contact
Michael Herrmann
Technical Marketing Non Rubber
Tel.: +49 (0)2 14-3 04 91 89
Email: michael.herrmann@lanxess.com

Further information
LANXESS is Germany's largest specialty
chemicals company and, with 15.000
employees, is present in 21 countries on
all the world's major markets. Listed on the
Frankfurt Stock Exchange, the company
achieved sales in 2008 of EUR 6.58 billion.

With its extensive portfolio, LANXESS
focuses on premium business. Its core
business comprises the development,
manufacture and sale of plastics, rubber,
specialty chemicals and intermediates.

It is also one of the world's leading suppliers
of synthetic rubber for the rubber-processing
industry.

The high-performance elastomer Levamelt®
(EVM), because of its range of special properties, is suitable for various applications
in adhesives technology.

Range of Products

Types of adhesives
Hot-melt adhesives
Solvent-based adhesives
Pressure-sensitive adhesives

Types of sealants
Various

Raw materials
Polymers:
Ethylene vinyl acetate copolymers

For applications in the field of
Paper/packaging
Bookbinding/graphics
Wood/furniture
Textiles
Adhesive tapes/labels
Hygiene

**Lohmannn
GmbH & Co. KG**
Irlicher Straße 55
D-56567 Neuwied
Phone +49 (0) 26 31-34-0
Fax +49 (0) 26 31-34-66 61
Email: info@lohmann-tapes.com
www.lohmann-tapes.com

Member of IVK

Company

Year of formation
1851

Size of workforce
approx. 1.400 worldwide

Managing directors:
Jürgen Walda
Manfred Meier
Dr. Norman Goldberg

Subsidiaries
I, F, E, PL, A, GB, DK, SE, RU, UA, China,
Korea, India and Mexico

Sales channels
Market segments:
Consumer Goods, Building & Construction,
Transportation, Mobile Communication,
Medical Technology and Hygiene

Contact partners
Marketmanaging and Marketing:
Ralph Unever

Further information
Lohmann offers mainly customized
adhesive solutions and takes care of
its customers from the first idea up
to automatic applications.
This is also shown in the new company logo:
Coating. Converting. Customizing.

Range of Products

Types of adhesives
Hot melt adhesives
Reactive adhesives
Solvent-based adhesives
Dispersion adhesives
Pressure-sensitive adhesives

Types of sealants
Acrylic sealants

Raw materials
Polymers

Equipment, plant and components for
Conveying, mixing, metering
Adhesive application
Surface pretreatment
Adhesive curing and drying
Measuring and testing

For applications in the field of
Paper/packaging
Bookbinding/plate mountig tapes
Wood/furniture industry
Construction industry, including floors,
walls and ceilings
Electronics
Automotive industry, aviation industry
Textile industry
Adhesive tapes, labels
Hygiene

Non-stop high technology.

When it comes to adhesive solutions for industrial applications, more and more customers are relying on Lohmann's experience and innovative abilities. This is no surprise, as our tapes not only offer perfect adhesive qualities but can also be equipped with numerous additional individual functions. Whether they are removable, vibration absorbent, transparent, electrical conductive or thermal conductive - the variety of applications is now almost limitless thanks to our expertise. And when will you receive our high-tech adhesive tapes?

LUGATO
GmbH & Co. KG

Großer Kamp 1
D-22885 Barsbüttel
Phone +49 (0) 40-6 94 07-0
Fax +49 (0) 40-6 94 07-1 10
Email: info@lugato.de
www.lugato.de

Member of IVK

Company

Year of formation
1919

Size of workforce
About 180

Ownership structure
GmbH & Co. KG, associated company of the Ardex GmbH since 2005

Sales channels
DIY stores

Contact partners
Application technology and sales:
info@lugato.de

Further information
www.lugato.de
www.lugato.com

Range of Products

Types of adhesives
Dispersion adhesives

Types of sealants
Acrylic sealants
Silicone sealants
MS/SMP sealants

For applications in the field of
Household, recreation and office

merz+benteli ag
Freiburgstraße 624
CH-3172 Niederwangen
Phone +41 (31) 980 48 48
Fax +41 (31) 980 48 49
Email: info@merz-benteli.ch
www.merz-benteli.ch

Company

Year of formation
1918

Size of workforce
85

Ownership structure
Familiy

Subsidiaries
none

Sales channels
direct & distributor

Contact partners
Management:
Paul Riedl, Marketing Director

Application technology and sales:
Mischa Thomi, Sales Manager Export

Range of Products

Types of adhesives
Reactive adhesives

Types of sealants
MS/SMP sealants

For applications in the field of
Wood/furniture industry
Construction industry, including floors,
walls and ceilings
Electronics
Mechanical engineering and equipment
construction
Automotive industry, aviation industry

Minova CarboTech GmbH
Am Technologiepark 1
D-45307 Essen
Phone +49 (0) 20 11 72-10 38
Fax +49 (0) 20 11 72-13 17
Email: info.de@minovaint.com
www.minova-ct.com

Member of IVK

Company

Size of workforce
125 employees at Essen (Minova Europe headquarters), 950 in Europe

Ownership structure
GmbH, Minova CarboTech is part of the Minova International Group which is member of Orica Ltd. (Australia)

Subsidiaries
Minova Europe has plants in Czech Republic, France, Germany, Great Britain, Iran, Italy, Kazakhstan, Poland, Russia, Spain, Sweden, Switzerland and Ukraine.

Contact partners
Management:
Dr. Jan Petzold (President and CEO Minova Europe), Detlev Beier (CFO), Dr. Archibald Richter, Marek Grocholewski

Application technology and sales:
Udo Nielbock, Mobile +49 (0) 1 72-2 66 37 41

Further information
Minova has more than 35 years worldwide experience in water sealing, strata consolidation, rock bolting and industrial adhesives. Minova is a leading supplier to mining, tunneling, geotechnical, civil engineering and adhesives industries. We offer a most complete and versatile portfolio for:
- Bolting systems for tunneling and civil engineering
- Resin systems for sealing, rock stabilisation and ground consolidation
- Accessories for bolting and injection work
- Adhesives for different kind of flooring

Range of Products

Types of adhesives
Reactive adhesives

Types of sealants
MS/SMP sealants

For applications in the field of
Wood/furniture industry
Construction industry, including floors, walls and ceilings

MORCHEM S.A.
Pol. Ind. Pla de Llerona
C/Alemania, 18 – 22
E-08520 Las Franquesas del Vallés
(Barcelona)
Phone +34 93 840 57 00
Fax +34 93 840 57 11
Email: morchem@morchem.com
www.morchem.com

Company

Year of formation
1985

Size of workforce
70 employees

Offices
Head office in Spain
Office Germany:
Morchem, S.A.
Schmuzerring 34
D-86609 Donauwörth
Phone +49 (0) 9 06-24 56 00
Fax +49 (0) 9 06-24 56 01
Email: ejochim@morchem.com

Warehouses
Barcelona/Spain
Frankfurt a. M./Central Europe
St. Petersburg, Moscow/Russia
Hongkong
Singapore
Egypt
Mexiko
El Salvador

Range of Products

Types of adhesives
Polyurethane adhesives:
Solvent based
Solvent free
Water based
Reactive hotmelts

Raw materials
Polyurethane dispersions

Applications
Polyurethane adhesives for:
Flexible packaging
Textile lamination
Graphic arts
Technical applications

Polyurethane dispersions for:
Coatings
Paints
Textile
Varnishes
Inks

Newsport Industries Ltd.

Chiswick
London, England
Phone +44 208 742 0333
Fax +44 208 742 0444
Email: sales@newport-industires.com
www.rosins.eu

Member of IVK

Company

Year of formation
1991

Size of workforce
14

Ownership structure
Private

Sales channels
Direct

Contact partners
Management:
Raj Patel

Application technology and sales:
Andrew Saunders

Further information
- Leading suppliers of rosins and resins to the adhesive industry
- Own office and technical centre in China
- European warehouse and stockist

Range of Products

Types of adhesives
Hot melt adhesives
Solvent-based adhesives
Dispersion adhesives

Types of sealants
Butyl sealants

Raw materials
Additives
Resins
Polymers

Novamelt-Jowat LLC, USA

Offinger Straße 120
D-79664 Wehr
Phone +49 (0) 77 82-70 85-0
Fax +49 (0) 77 82-70 85-25
Email: info@novamelt.de
www.novamelt.com

Member of IVK

Company

Year of formation
1989

Size of workforce
50

Managing partners
Harald Braun
Pieter W. Mol

Subsidiaries
Novamelt-Jowat LLC, USA

Contact partners
Management:
Harald Braun

Application technology and sales:
Peter Hölzle

Further information
Specialisation in hot melt pressure sensitive
adhesives

Range of Products

Types of adhesives
Hot melt adhesives
Pressure-sensitive adhesives

For applications in the field of
Paper/packaging
Construction industry, including floors,
walls and ceilings
Mechanical engineering and equipment
construction
Automotive industry
Textile industry
Adhesive tapes, labels

Omya Peralta GmbH

Kanalstraße 44
D-22085 Hamburg
Phone +49 (0) 40-6 58 72-0
Email: info@omya-peralta.de
www.omya-peralta.de

Member of IVK

Company	Range of Products

Contact
Application technology & sales:
Sven.Wiemer@Omya-Peralta.de

Raw materials
Solvents:
Triacetin-food grade (plasticizer), process
and mineral oils
Polymers:
Polyvinylalcohol (PVA), Polyisobutylen (PIB),
Polyvinylpyrrolidone (PVP), Cellulosic Ether
(HPMC, HEMC – modified, unmodified),
Ethylene Vinyl ectate (EVA)

For applications in the field of
Paper/packaging
Bookbinding/graphic design
Wood/furniture industry
Construction industry, including floors,
walls and ceilings
Textile industry
Household, recreation and office

Organik Kimya Netherlands B. V.

Chemieweg 7, Havenummer 4206
NL-3197KC Rotterdam Botlek,
Phone +31 10 295 48 20
Fax +31 10 295 48 29
Email: organik@organikkimya.com
www.organikkimya.com

Member of IVK

Company

Year of formation
1924

Size of workforce
350

Managing partners
Simone Kaslowski
Stefano Kaslowski

Nominal capital
100 %

Ownership structure
100 % familiy owned business

Subsidiaries
Distribution worldwide, production sites in
the Netherlands and Turkey

Sales channels
Direct sales and through distributors
worldwide

Contact partners
Management:
Stefano Kaslowski, General Manager

Application technology and sales:
Oguz Kocak, Sales Manager
Phone: +49 (0) 173-6 52 22 59
Email: o_kocak@organikkimya.com

Range of Products

Types of adhesives
Dispersion adhesives
Pressure-sensitive adhesives

Types of sealants
Acrylic sealants

Raw materials
Polymers

For applications in the field of
Paper/packaging
Bookbinding/graphic design
Wood/furniture industry
Construction industry, including floors,
walls and ceilings
Automotive industry, aviation industry
Textile industry
Adhesive tapes, labels

Hermann Otto GmbH
Krankenhausstraße 14
D-83413 Fridolfing
Phone +49 (0) 86 84-9 08-0
Fax +49 (0) 86 84-9 08-5 39
Email: info@otto-chemie.com
www.otto-chemie.com

Member of IVK

Company

Management board
Jürgen Lohre
Dr. Volker Weidmann
Johann Hafner
Matthias Nath

Size of workforce
300 employees

Contact person
Application support:
Nikolaus Auer
Phone +49 (0) 8 68 49 08-4 56
Email: nikolaus.auer@otto-chemie.de

Sales manager industrial applications:
Ralf Burzlaff
Phone +49 (0) 8 68 49 08-5 28
Email: ralf.burzlaff@otto-chemie.de

Range of Products

Types of sealants and adhesives
1-comp. and 2-comp. silicones
1-comp. and 2-comp. polyurethanes
1-comp. and 2-comp. hybrids
acrylates

For applications in the field of
Vehicle manufacturing, rail, shipbuilding,
caravan, superstructures,
aeronautics,
electro technology, household appliances,
hobs, backing ovens,
electronics and cable industry,
glass industry,
manufacturing of aquaria,
weather sealing, fire protection,
metal, windows and doorframes,
container construction,
heating, ventilation, and air conditioning,
wood, furniture, sandwich elements,
flooring
plastics processing
clean rooms
photovoltaic modules
hot water modules
partition walls
manufacturing of prefab houses and
garages
textile coating

Panacol-Elosol GmbH

Daimler Straße 8
D-61449 Oberursel
Phone +49 (0) 61 71 62 02-0
Fax +49 (0) 61 71 62 02-5 90
Email: info@panacol.de
www.panacol.de

Member of IVK

Company

Year of formation
1978

Size of workforce
50

Managing partners
Achim Battermann
Florian Eulenhöfer

Nominal capital
255.645 €

Ownership structure
Panacol-Elosol GmbH is the German subsidiary of Panacol AG and Elosol AG in Switzerland. Panacol Group is the subsidiary of Dr. Hönle AG

Sales channels
Sales team for Germany
International distribution network with worldwide sales partners

Contact partners
Management: Achim Battermann
Application technology: Jürgen Perl
Application sales: Joachim Becker

Further information
Panacol offers a wide range of products, ranging from UV adhesives through structural adhesives to silicones. Furthermore is Panacol a system provider and distributes a wide range of dispensing equipment and machines, which match exactly with the adhesives and applications. Since 1 January 2008 Panacol has been a member of the Hönle Group. Since February 2009 Panacol is responsible for the Group entire distribution line of adhesives, coatings and potting materials.

Range of Products

Types of adhesives
Reactive adhesives
Anaerobic adhesives
Cyanacrylate
UVA/visible light curing epoxy and acrylate adhesives
High temperature adhesive
Conductive isotropic and anisotropic adhesives
Momentive Performance Materials Silicones
Epoxy resins moulding
Structural adhesive

Types of sealants
Acrylic sealants
Silicone sealants
Other Epoxies

Equipment, plant and components
for conveying, mixing, metering and
for adhesive application
for adhesive curing

For applications in the field of
Electronics and electro-technics
Mechanical engineering and equipment construction
Automotive industry, aviation industry
Adhesive tapes, labels
Conformal coating
Glass converting and finishing industry
Household and office
Optics
Timber and furniture industry
Medical technology

PCI Augsburg GmbH

Piccardstraße 11
D-86159 Augsburg
Phone +49 (0) 8 21 59 01-0
Fax +49 (0) 8 21 59 01-3 72
Email: pci-info@basf.com
www.pci-augsburg.com

Member of IVK

Company

Year of formation
1950

Size of workforce
800

Ownership structure
PCI Augsburg GmbH is part of BASF –
The Chemical Company

Subsidiaries
please refer to website for details

Sales channels
indirect/through distributors

Contact partners
Management:
please refer to website

Application technology and sales:
please refer to website

Further information
please refer to website

Range of Products

Types of adhesives
Reactive adhesives
Dispersion adhesives

Types of sealants
Acrylic sealants
PUR sealants
Silicone sealants

Equipment, plant and components
or conveying, mixing, metering and
for adhesive application

For applications in the field of
Construction industry, including floors,
walls and ceilings

PLANATOL Adhesive GmbH
Fabrikstraße 30 – 32
D-83101 Rohrdorf
Phone +49 (0) 80 31-7 20-0
Fax +49 (0) 80 31-7 20-1 80
Email: info@planatol.de
www.planatol.de

Member of IVK

Company

Year of formation
1932

Size of workforce
60

Ownership structure
PLANATOL Klebtechnik GmbH

Sales channels
Direct sale to industrial customers
Own sales staff and sales representatives
Graphic retailers
World-wide sales and service network

Contact partners
Managing director:
Dr. Jörg Dietrich
Phone +49 (0) 80 31-7 20-1 54

Sales Manager:
Dr. Jörg Dietrich
Phone +49 (0) 80 31-7 20-1 54

Application technique and development:
Dr. Rainer Fasolt
Phone +49 (0) 80 31-7 20-1 37

Further information
PLANATOL is one of the leading suppliers of
adhesives but the company also produces
OPIMATIC systems for rotary fold gluing
in web presses, SOFTJET systems for fold
softening, PLANAX perfect binding devices
and bookbinding materials for bookbinders,
copy-shops, in-house printing departments
or offices and high functional adhesive tapes.

Range of Products

Types of adhesives
Hot melt adhesives
Reactive adhesives
Solvent-based adhesives
Dispersion adhesives
Vegetable adhesives, dextrin and
starch adhesives
Pressure-sensitive adhesives

For applications in the field of
Paper/packaging
Bookbinding/graphic design
Construction industry, including floors,
walls and ceilings
Automotive industry, aviation industry
Adhesive tapes, labels

Polimeri Europa SpA.

Piazza Boldrini, 1
San Donato Milanese
Email: info@polimerieuropa.com
www.polimerieuropa.com

Member of IVK through Polimeri Europa
GmbH, Eschborn, Germany

Company

Year of formation
1957

Size of workforce
6.500

Managing partners
Ing. Antonio Protopapa,
Commercial Director Polymers

Nominal capital
1.553.400.000

Ownership structure
ENI SpA.

Subsidiaries
see website

Sales channels
Polimeri Europa Sales Network in major
European countries, see website

Contact partners
Management:
Giovanni Cassuti,
Sales Manager Polymers + local staff
in single countries

Further information
See official website:
www.polimerieuropa.com

Range of Products

Types of adhesives
Hot melt adhesives
Solvent-based adhesives
Pressure-sensitive adhesives

Raw materials
Polymers

For applications in the field of
Paper/packaging
Bookbinding/graphic design
Wood/furniture industry
Automotive industry, aviation industry
Adhesive tapes, labels
Hygiene
Household, recreation and office

PolymerLatex GmbH
Werrastraße 10
D-45768 Marl
Phone +49 (0) 23 65 49 25 52
Email: info@polymerlatex.de
www.polymerlatex.com

Member of IVK

Company

Year of formation
1996

Size of workforce
~ 600 worlwide

Contact partners
Management:
Dr. Martin Conrads

Range of Products

Types of adhesives
Dispersion adhesives
Pressure-sensitive adhesives

Types of sealants
Acrylic sealants
Other

Raw materials
Additives:
Thickener
Polymers:
Acrylics, XSBR, HS-SBR and CR

For applications in the field of
Paper/packaging
Bookbinding/graphic design
Wood/furniture industry
Construction industry, including floors,
walls and ceilings
Automotive industry, aviation industry
Textile industry
Adhesive tapes, labels
Hygiene
Household, recreation and office

PRHO-CHEM GmbH
Dohlenstraße 8
D-83101 Rohrdorf-Thansau
Phone +49 (0) 80 31-3 54 92-0
Fax +49 (0) 80 31-3 54 92-29
Email: info@prho-chem.de
www.prho-chem.de

Member of IVK

Company

Year of formation
1994

Ownership structure
private property

Contact partners
Management:
Otto Kleinhanß

Application technology and sales:
Otto Kleinhanß

Range of Products

Types of adhesives
Hot melt adhesives
Reactive adhesives
Dispersion adhesives
Glutine glue
Pressure-sensitive adhesives

Types of sealants
PUR sealants
Silicone sealants
MS/SMP sealants

For applications in the field of
Paper/packaging
Bookbinding/graphic design
Automotive industry, aviation industry
Hygiene

Renia
Gesellschaft mbH

Ostmerheimer Straße 516
D-51109 Köln
Phone +49 (0) 2 21-63 07 99-0
Fax +49 (0) 2 21-63 07 99-50
Email: info@renia.com
www.renia.com

Member of IVK

Company

Year of formation
1930

Ownership structure
GmbH, family owned

Sales channels
exclusive partners in more than 50 countries

Contact partners
Management:
Heinz Buchholz

Application technology and sales:
Dr. Julian Grimme

Range of Products

Types of adhesives
Solvent-based adhesives
Dispersion adhesives

For applications in the field of
Household, recreation and office
Shoe Industry
Health

Rhenocoll-Werk e. K.

Erlenhöhe 20
D-66871 Konken
Phone +49 (0) 63 84-99 38-0
Fax +49 (0) 63 84-99 38-1 12
Email: info@rhenocoll.de
www.rhenocoll.de

Member of IVK

Company

Year of formation
1948

Size of workforce
120

Ownership structure
Joint partnership, Fam. Holding

Subsidiaries
Polska, Czechoslovakia, Belarus, Georgia,
Russia, China

Sales channels
Dealer based worldwide

Contact partners
Management:
Werner Zimmermann

Range of Products

Types of adhesives
Hot melt adhesives
Reactive adhesives
Dispersion adhesives
Pressure-sensitive adhesives

Types of sealants
Acrylic sealants
PUR sealants

For applications in the field of
Paper/packaging
Wood/furniture industry
Construction industry, including floors,
walls and ceilings
Household, recreation and office

RÜTGERS
Germany GmbH

Varziner Straße 49
D-47138 Duisburg
Phone +49 (0) 2 03-42 96-02
Fax +49 (0) 2 03-42 96-7 62
Email: harze@ruetgers-group.com
www.novares.de

Member of IVK

Company

Year of formation
1849

Sales channels
own sales force

Contact partners
Management:
Uwe Holland, Head of Business Line
Performance Products

Application technology and sales:
Claudia Hinz, Manager Application
Technology Adhesive Industry

Range of Products

Raw materials
Hydrocarbon resins
Resins: aromatic HCR,
aliphatically mod. HCR,
Indene-coumarone resins,
phenolic mod. HCR,
Special liquid resins

SABA
Dinxperlo BV

Industriestraat 3
NL-7091 DC Dinxperlo
Phone +31 315 65 89 99
Fax +31 315 65 32 07
Email: info@saba.nl
www.saba.nl

Member of IVK

Company

Year of formation
1933

Size of workforce
140

Managing director
C. P. M. Lansbergen

Ownership structure
R. J. Baruch, H. G. Otten

Subsidiaries
SABA Polska SP. z o. o.
SABA North America LLC
SABA Bocholt GmbH
SABA Nordica

Contact:
Business unit industry: industry@saba.nl
Business unit building & construction:
info@saba.nl

Further information
Development, production and sales of high-quality adhesives and sealants for industrial and constructional applications.
Business unit industry. Adhesives for:
• foam processing industry: mattress and furniture production
• pvc industry: horticulture, industry and pool construction
• insulation bonding and interior decorating

Business unit building &construction.
Sealants for:
• marine
• transportation
• building & construction
• civil & environmental engineering

Range of Products

Types of adhesives
Hot melt adhesives
Reactive adhesives
Solvent-based adhesives
Dispersion adhesives
Pressure-sensitive adhesives

Types of sealants
Polysulfide sealants
PUR sealants
Silicone sealants
MS/SMP sealants

Equipment, plant and components
for conveying, mixing, metering and
for adhesive application
for surface pretreatment

For applications in the field of
Wood/furniture industry
Construction industry, including floors, walls and ceilings
Automotive industry, aviation industry
Textile industry
Household, recreation and office

SCHÖNOX GmbH

Alfred-Nobel-Straße 6
D-48720 Rosendahl
Phone +49 (0) 25 47-9 10-0
Fax +49 (0) 25 47-9 10-1 01
Email: info@schoenox.de
www.schoenox.com

Member of IVK

Company

Year of formation
1891

Size of workforce
320

Ownership structure
100 % subsidiary of Akzo Nobel Farben Beteiligungs-GmbH

Sales channels
Wholesalers for tiling, floor coverings and general building materials

Contact partners
Management:
Helmut Twilfer, General Manager Sub Business Unit Building Adhesives

Application technology and sales:
Werner Schwerdt, Sales & Marketing Director Germany, Austria, Switzerland and Benelux

Range of Products

Types of adhesives
Reactive adhesives
Solvent-based adhesives
Dispersion adhesives

Types of sealants
Acrylic sealants
PUR sealants
Silicone sealants
MS/SMP sealants

For applications in the field of
Construction industry, including floors, walls and ceilings

Schomburg GmbH & Co. KG

Aquafinstraße 2 - 8
D-32760 Detmold
Phone +49 (0) 52 31-9 53-00
Fax +49 (0) 52 31-9 53-1 23
Email: info@schomburg.de
www.schomburg.de

Member of IVK

Company

Year of formation
1937

Size of workforce
220 (Germany), 520 worldwide

Managing partners
Albert Schomburg
Ralph Schomburg
Dr. Jens Hofele

Nominal capital
3.438 Mio. €

Ownership structure
Family and Management owned

Subsidiaries
30 worlwide:
Poland, Czech Republic, USA, India, Turkey,
Luxemburg, Switzerland, Russia, Netherlands, Slovakia, etc.

Sales channels
Distribution partners

Contact partners
Management:
Dr. Jens Hofele

Application technology and sales:
Holger Sass
Michael Hölscher

Range of Products

Types of adhesives
Reactive adhesives
Dispersion adhesives
Cement based adhesives

Types of sealants
Acrylic sealants
Polysulfide sealants
PUR sealants
Silicone sealants
Other

Equipment, plant and components
for conveying, mixing, metering and
for adhesive application

For applications in the field of
Construction industry, including floors,
walls and ceilings
Mechanical engineering and equipment
construction

the plus of pure
performance

Schülke & Mayr GmbH
Robert-Koch-Straße 2
D-22851 Norderstedt
Phone +49 (0) 40-5 21 00-0
Fax +49 (0) 40-5 21 00-2 44
Email: sai@schuelke.com
www.schuelke.com

Member of IVK

Company

Foundation
1889

Manpower
580 employees

Shareholder
Air Liquide, Paris

Distribution channels
Own sales representatives

Subsidiaries
UK, Netherlands, Italy, Switzerland,
France, Shanghai and distributors worldwide

Contact person
Market Manager (technical biocides):
Ingo Krull

Further information
Production and distribution of chemical
and technical biocides as well as hygiene
products for cleaning and disinfection

Range of Products

Raw materials
Additives:
Preservatives, Biocides
Other:
Microbiological Services

For applications in the field of
Paper/packaging
Wood/furniture industry
Building industry incl. electronics
Engineering industry and construction of
apparatus
Vehicles
Textile industry
Adhesive tapes
Hygiene
Household, hobby and office

Sika Automotive GmbH
Reichsbahnstraße 99
D-22525 Hamburg
Phone: +49 (0) 40-5 40 02-0
Fax: +49 (0) 40-5 40 02-5 15
Email: info.automotive@de.sika.com
www.sika-automotive.de

Member of IVK

Company

Year of formation
1928

Size of workforce
170

Subsidiaries
Sister companies in 75 countries

Contact partners
Managing Director:
Dirk Brandenburger

Sales Germany:
Dirk Brandenburger
Andreas Finkenzeller

Export:
Dirk Brandenburger

Range of Products

Types of adhesives
Hot melt adhesives
Reactive adhesives
Solvent-based adhesives
Dispersion adhesives
Pressure-sensitive adhesives

Types of sealants
PUR sealants
Other

For applications in the field of
Electronics
Automotive industry, aviation industry
Textile industry
Adhesive tapes & labels
Hygiene

**Creating Solutions
for Increased Productivity**
Sika is supplier and development partner to
the automotive industry. Our state-of-the-art
technologies provide solutions for increased
structural performance, added acoustic
comfort and improved production processes. As a specialty company for chemical
products, we concentrate on our core
competencies: **Bonding – Sealing –
Damping – Reinforcing**
As a globally operating company, we are
partner to our customers worldwide. Sika
is represented with its own subsidiaries in
all automobile-producing countries, thus
guaranteeing a professional and fast local
service.

Automotive

Promotional Supplier of th
BMW Sauber F1 Tear

Sika Automotive
Solutions for Bonding, Sealing,
Damping and Reinforcing.

www.sika-automotive.de

Sika Deutschland GmbH

Stuttgarter Straße 139
D-72574 Bad Urach
Phone +49 (0) 71 25 9 40 7 61
Fax +49 (0) 71 25 9 40 7 61
Email: industry@de.sika.com
www.sika.com

Member of IVK

Company

Year of formation
1910

Size of workforce
12.900 (worldwide Sika AG),
1.400 Sika Deutschland

Subsidiaries
in more than 700 countries,
see www.sika.com

Sales channels
direct and distribution

Range of Products

Types of adhesives
Hot melt adhesives
Reactive adhesives
Solvent-based adhesives
Dispersion adhesives
Pressure-sensitive adhesives
PUR adhesives

Types of sealants
Acrylic sealants
Butyl sealants
Polysulfide sealants
PUR sealants
Silicone sealants
MS / SMP sealants
Other

For applications in the field of
Wood / furniture industry
Construction industry, including floors,
walls and ceilings
Electronics
Mechanical engineering and equipment
construction
Automotive industry, aviation industry
Adhesive tapes, labels
Household, recreation and office

Stop looking. Start finding.

Information that inspires is the root of innovation. Staying up to date helps accelerate development. And substance is what makes knowledge valuable. ATZonline is the place to go when you want to know what's happening in our industry and to get information that is unique in its depth. ATZ, MTZ, ATZelektronik, ATZproduktion and ATZautotechnology subscribers get access to a complimentary archive of industry articles as well as specials and whitepapers. All articles are well researched, with background and insider information.

No need to look any further – get your competitive advantage on **www.ATZonline.com**

ATZonline. Know more. Go further.

ATZ online

Springer Automotive Media subscription service | Abraham-Lincoln-Str. 46 | D-65189 Wiesbaden
Phone: +49 611.7878-151 | Fax: +49 611.7878-423 | SpringerAutomotive@abo-service.info | www.ATZonline.com

Sonderhoff Engineering GmbH

Allgäustraße 3
A-6912 Hörbranz
Phone +43 55 73-8 29 91
Fax +43 55 73-8 29 46
Email: info@sonderhoff.com
www.sonderhoff.com

Company

Year of formation
1988

Size of workforce
90 employee
(Sonderhoff group 180)

Ownership structure
Sonderhoff Holding GmbH

Subsidiaries
Sonderhoff Chemicals GmbH, Cologne
Sonderhoff Services GmbH, Cologne
Sonderhoff Iberica S.A.U., Esparraguera (Spain)
Sonderhoff Italia s.r.l., Valmadrera (Italy)
Sonderhoff USA Corp., Elgin (USA)
Sonderhoff (Suzhou) Sealing Systems Co. Ltd.,
Suzhou (China)

Sales channels
worldwide

Contact partners
Application technology and sales:
Jörg Kessler

Further information
We develop dosing machines and automation
concepts according to your specifications,
from stand-alone systems up to fully-auto-
mated production lines, as well as gasketing,
gluing and potting.

Range of Products

Types of adhesives
Reactive adhesives

Types of sealants
PUR sealants
Silicone sealants
PVC

Raw materials
Polymers

Equipment, plant and components
for conveying, mixing, metering and
for adhesive application

For applications in the field of
Paper/packaging
Electronics
Mechanical engineering and equipment
construction
Automotive industry, aviation industry

Stauf Klebstoffwerk GmbH
Oberhausener Straße 1
D-57234 Wilnsdorf
Phone +49 (0) 27 39-3 01-0
Fax +49 (0) 27 39-3 01-2 00
Email: info@stauf.de
www.stauf.info

Member of IVK

Company

Year of formation
1828

Size of workforce
55

Ownership stucture
100 % Family Stauf

Managerial head
Volker Stauf, Wolfgang Stauf

Sales channels
Worldwide distribution, own field service
and distributing warehouses in Germany
and other countries for:
• the wood flooring wholesale
• the floor covering wholesale
• the construction material wholesale
• handcraft enterprises
• contractors
• architects

Products
• adhesive systems for floor covering and
  wood flooring
• mounting repair adhesives
• artificial turf adhesives
• primers
• levelling compounds
• underlayments
• accessories

Contact partners
Management:
Volker Stauf, Phone +49 (0) 27 39-3 01-1 28,
v.stauf@stauf.de

Product Technology:
Dr. Frank Gahlmann, Phone +49 (0) 27 39-3
01-1 65, gahlmann@stauf.de

Range of Products

Types of adhesives
Reactive adhesives
Solvent-based adhesives
Dispersion adhesives
Powder adhesives

For applications in the field of
Construction industry, including floors,
walls and ceilings
• Parquet and wood flooring
• End grain wood blocks
• Textile and elastic floor coverings
• Mounting repair
• Industrial application

Further information
www.stauf.info

STAUF is a leading system supplier for flooring technology. For the safe and durable
bonding of wood flooring and floor coverings we research, develop and produce
innovative adhesive systems on a high-
grade raw material basis. Next to the latest
adhesive systems STAUF offers the whole
bandwidth of products for sub floor preparation and accessories as well.

STAUF has remained a family-owned
company even after more than 180 years.

Long time experience as well as continuous
advancement in a state of the art production and research environment ensure the
constant top-level product quality and set
the standards for the customers of the
wood flooring and floor covering branch

Synthopol Chemie

Alter Postweg 35
D-21614 Buxtehude
Phone +49 (0) 41 61-70 71 49
Fax +49 (0) 41 61-8 01 30
Email: bprueter@synthopol.com
www.synthopol.com

Member of IVK

Company

Year of formation
1957

Size of workforce
190

Ownership structure
Family company

Sales channels
Germany and Europe

Contact partners
Management:
Dr. Rüdiger Spohnholz
Phone +49 (0) 41 61-70 71 25

Sales:
Hubert Starzonek (Commercial Manager)
Phone +49 (0) 41 61-70 71 32

Application technology:
Dr. Rüdiger Spohnholz
Phone +49 (0) 41 61-70 71 25

Further information
Birgit Prüter
Phone +49 (0) 41 61-70 71 49

Range of Products

Types of adhesives
Hot melt adhesives
Reactive adhesives
Solvent-based adhesives
Dispersion adhesives
Pressure-sensitive adhesives

Types of sealants
Acrylic sealants

Raw materials
Resins
Polymers

For applications in the field of
Construction industry, including floors,
walls and ceilings
Mechanical engineering and equipment
construction
Automotive industry, aviation industry
Textile industry
Adhesive tapes, labels

TER HELL & Co. GmbH

Börsenbrücke 2
D-20457 Hamburg
Phone +49 (0) 40-30 05 01-0
Fax +49 (0) 40-33 50 50
Email: info@terhell.com
www.tergroup.com

Member of IVK

Company

Year of formation
1908

Size of workforce
439

Managing partners
Klaus C. Westphal
Christian Westphal

Nominal capital
45 Mio.

Subsidiaries
36

Sales channels
Trading, Distribution, Salesforce

Contact partners
Management:
Wolfgang Weger
Christian Westphal

Application technology and sales:
Uwe Lüdemann
Jens Vinke

Further information
www.terhell.com

Range of Products

Types of adhesives
Hot melt adhesives
Reactive adhesives
Solvent-based adhesives
Dispersion adhesives
Pressure-sensitive adhesives

Types of sealants
Butyl sealants
PUR sealants
Pib

Raw materials
Additives
Fillers
Resins
Solvents
Polymers

For applications in the field of
Paper/packaging
Bookbinding/graphic design
Wood/furniture industry
Construction industry, including floors,
walls and ceilings
Electronics
Automotive industry, aviation industry
Textile industry
Adhesive tapes, labels
Hygiene

Best and worst practices in sales and customer management

Dannenberg, Holger | Zupancic, Dirk
Excellence in Sales
Optimising Customer and Sales Management
2009. xxxii, 232 pp. With 71 Fig. and 12 Tab.
Hardc. EUR 49,90
ISBN 978-3-8349-1006-6

„Excellence in Sales" is an integrated management approach for professional sales organisations. The authors collected best and worst practices in sales and customer management. The concept for true excellence in sales is relying on a set of levers which are explained in a systematic manner. Readers get the chance to compare their solutions with the worldwide top performers. Illustrations and numerous recommendations for implementation show how to improve the overall performance of companies.

The Authors
Holger Dannenberg is Managing Partner of the Executive Board of Mercuri International Deutschland GmbH.
Dr. Dirk Zupancic is Professor of Management at Heilbronn Business School, Germany, senior lecturer of business administration and head of the competence centre BtoB Marketing and Sales at the University of St. Gallen, Switzerland.

tesa SE
Quickbornstraße 24
D-20253 Hamburg
Phone +49 (0) 40-49 09-1 01
Fax +49 (0) 40-49 09-60 60
www.tesa.de
www.tesa.com

Member of IVK

Company

Year of formation
tesa AG 2001,
tesa SE since march, 30th 2009

Size of workforce
3,900

Ownership structure
100 % subsidiary of Beiersdorf AG, Hamburg

Subsidiaries
51

Sales channels
Industry (e. g. automotive, electronics,
print & paper, solar), Food and DIY

Contact
www.tesa.com

Further information
tesa SE is one of the world's leading
manufacturers of technical adhesive tapes
and self-adhesive system solutions. The
company's focus is on innovative problem-
solving for industrial customers and consumers. The spectrum of applications in the
industrial field ranges from special adhesive
tapes for the printing and paper industries
via cable loom tapes for cars and high-
performance products for fixing electronic
components in mobile phones and digital
cameras to forgery-proof laser labels.
tesa became world famous for branded
products for end consumers – like tesa

Range of Products

Types of adhesives
Hot melt adhesives tapes
Reactive adhesives tapes
Pressure-sensitive adhesives tapes

For applications in the field of
Paper/packaging
Bookbinding/graphic design
Wood/furniture industry
Construction industry, including floors,
walls and ceilings
Electronics
Mechanical engineering and equipment
construction
Automotive industry, aviation industry
Adhesive tapes, labels
Household, recreation and office

dictionary. 300 professional aids are available
to consumers at DIY and discount stores
for creating their personal environment at
work, at home or in the garden.
tesa SE is a member of the Beiersdorf
Group and has been an independent public
company since 2001. With a total staff of
3,900 in 51 affiliates, tesa is active in more
than 100 countries worldwide.

tesa SE
A Beiersdorf Company

Türmerleim GmbH

Arnulfstraße 43
D-67061 Ludwigshafen
Phone +49 (0) 6 21-5 61 07-0
Fax +49 (0) 6 21-5 61 07-12
Email: info@tuermerleim.de
www.tuermerleim.de

Member of IVK

Company

Year of formation
1889

Size of workforce
130

Managing partners
Dr. Hermann Pfeiffer
Matthias Pfeiffer
Dr. Thomas Pfeiffer

Nominal capital
3.100.000 €

Ownership structure
Family company

Subsidiaries
Türmerleim AG, Basel

Contact partners
Management:
Dr. Jörg Liebe
Josef Karl
Tanguy Trippner
Harald Staub

Application technology and sales:
see Management

Range of Products

Types of adhesives
Hotmelts
Emulsions
Starch, dextrin and casein adhesives
UF-/MUF-resins

For applications in the field of
Paper/packaging
Labelling
Wood/furniture industry
Hygiene

Uzin Utz AG

Dieselstraße 3
D-89079 Ulm
Phone +49 (0) 7 31-40 97-0
Fax +49 (0) 7 31-40 97-1 10
Email: info@uzin.com
www.uzin.com

Member of IVK

Company

Year of formation
1911

Size of workforce
810 (May 2009)

Managing partners
Dr. H. Werner Utz
Thomas Müllerschön

Ownership structure
Public company

Subsidiaries
30 all over the world

Sales channels
Trade and distribution

Contact partners
Management:
Mr. Ludin

Application technology and sales:
Mr. Ludin

Further information
For more than 60 years, Uzin has been a
term for professional installation systems
for flooring and wood flooring.

Range of Products

Types of adhesives
Reactive adhesives
Solvent-based adhesives
Dispersion adhesives
Pressure-sensitive adhesives
Adhesives on PE film carrier

Types of sealants
Acrylic sealants
PUR sealants
Silicone sealants

Equipment, plant and components
for conveying, mixing, metering and
for adhesive application
for surface pretreatment
measuring and testing

For applications in the field of
Construction industry, including floors,
walls and ceilings

Vinavil S. p. A.

Via Valtellina, 63
I-20159 Milano
Phone +39-02-69 55 41
Fax +39-02-69 55 48 90
Email: vinavil@vinavil.it
www.vinavil.it

Member of IVK

Company

Year of formation
1994

Size of workforce
> 300

Ownership structure
Mapei S. p. A.

Subsidiaries
Vinavil Americas. Corp.
Vinavil Egypt
Vinavil Far East Ltd.

Sales channels
> 40 representative commercial offices

Contact partners
Management:
Dr. Zaverio Rovea
Silvio Pellerani

Application technology and sales:
Silvio Pellerani
Dr. Fabio Chiozza

Further information
Certified acc. ISO EN 9001 and 14001

Range of Products

Types of adhesives
Dispersion adhesives
Pressure-sensitive adhesives
Solvent-adhesives

Types of sealants
Acrylic sealants

Raw materials
Polymers:
aqueous polymer dispersions,
solid resins and redispersible powders
RAVEMUL®, VINAVIL®, CRILAT®, RAVIFLEX®
and VINAFLEX® based on vinylacetate, vinylacetate copolymers, vinylacetate ethylene
copolymers, acrylic and styrene acrylic

For applications in the field of
Paper/packaging
Bookbinding/graphic design
Wood/furniture industry
Construction industry, including floors,
walls and ceilings
Automotive industry, aviation industry
Textile industry
Adhesive tapes, labels
Household, recreation and office

VITO IRMEN GmbH & Co. KG
Mittelstraße 74 – 80
D-53424 Remagen
Phone +49 (0) 26 42-4 00 70
Fax +49 (0) 26 42-4 29 13
Email: info@vito-irmen.de
www.vito-irmen.de

Member of IVK

Company

Year of formation
1907

Size of workforce
86

Managing director
Ralf Heiligtag

Nominal capital
4.000.000 €

Ownership structure
Limited commercial partnership

Subsidiaries
in Poland and Austria

Sales channels
direct and via dealers & distributors

Contact partners
Management:
Ralf Heiligtag

Application technology and sales:
Andreas Hollweg

Further information
www.vito-irmen.de

Range of Products

Types of adhesives
Hot melt adhesives
Solvent-based adhesives
Dispersion adhesives
Pressure-sensitive adhesives

For applications in the field of
Wood/furniture industry
Construction industry, including floors,
walls and ceilings
Mechanical engineering and equipment
construction
Automotive industry, aviation industry
Adhesive tapes, labels
Hygiene

Wacker Chemie AG
Hanns-Seidel-Platz 4
D-81737 München
Phone +49 (0) 89-62 79-0
Fax +49 (0) 89-62 79-17 70
Email: info@wacker.com
www.wacker.com

Member of IVK

Company

Year of foundation
1914

Size of workforce
about 16.000 (2008)

Ownership structure
Stock corporation ("Aktiengesellschaft")

Subsidiaries
27 production sites and some 100
subsidiaries and offices wordwide.

Range of Products

Raw Materials
Polymers:
Vinyl acetate polymers
dispersions, dispersible polymer powders
and solid resins (VINNAPAS® and VINNOL®)
Vinyl acetate/ethylene copolymers
dispersions and dispersible polymer pow-
ders (VINNAPAS®)
VC copolymers (VINNAPAS®)
Polyvinyl butyrals
Silicones

Additives:
Pyrogenic silica (HDK®)
Silanes, adhesion promoters and cross-
linkers (GENIOSIL®)
Foam-control agents and silicone
surfactants
Nanoscale silicone particles for modifying
adhesives (GENIOPERL®)

Sealants and Adhesive Grades
RTV-1 Silicones
RTV-2 Silicones
LSR Silicones
Silicone gels and silicone foams
UV-curing systems
Hybrid adhesives

Wakol GmbH
Bottenbacher Straße 30
D-66954 Pirmasens
Phone +49 (0) 63 31-80 01-0
Email: info@wakol.com
www.wakol.com

Member of IVK

Company

Year of formation
1934

Size of workforce
150

Management
Heinz-Dieter Groß
Dr. Götz Hillert
Christian Groß

Subsidiaries
Austria, Switzerland, Poland, Italy, USA

Sales channels
direct distribution, specialised trade

Range of Products

Types of adhesives
Hot melt adhesives
Reactive adhesives
Solvent-based adhesives
Dispersion adhesives

For applications in the field of
Construction industry, including floors,
walls and ceilings
Automotive industry, aviation industry
Textile, wood/furniture industry

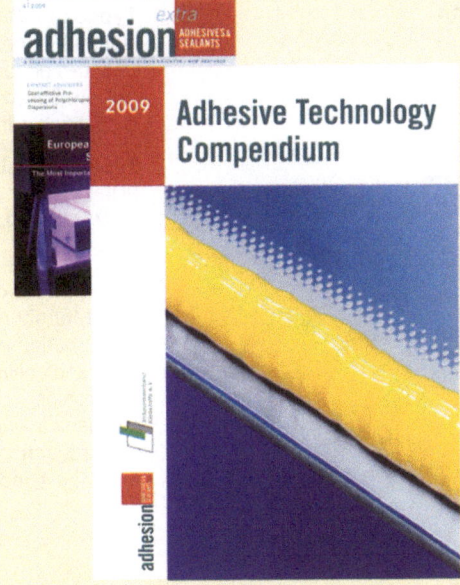

Weiss Chemie + Technik GmbH & Co. KG

Hansastraße 2
D-35708 Haiger
Phone +49 (0) 27 73-8 15-0
Fax +49 (0) 27 73-8 15-2 00
Email: ch@weiss-chemie.de
www.weiss-chemie.de

Member of IVK

Company

Year of formation
1815

Size of workforce
260 members of staff within the company group

Managing partners
WBV-Weiss Beteiligungs- und Verwaltungs-gesellschaft mbH

Nominal capital
2 Mio. €

Ownership structure
Family share holders

Subsidiaries
Haiger, Herzebrock, Niederdreisbach, Monroe North Carolina (USA)

Sales channels
Specialist wholesaler, industry and trade

Contact partners
Management:
Jürgen Grimm

Sales: Phone +49 (0) 27 73-8 15-2 11
Technology: Phone +49 (0) 27 73-8 15-2 55
Purchase: Phone +49 (0) 27 73-8 15-2 41

Further information
Division sandwich elements
Light sandwich constructions as heat- and sound insulating sandwich elements applied in fields like doors, windows, gates, booth constructions, automotive industry, etc

Range of Products

Types of adhesives
Hot melt adhesives
Reactive adhesives
Solvent-based adhesives
Dispersion adhesives

Types of sealants
PUR sealants
MS/SMP sealants
Other

For applications in the field of
Wood/furniture industry
Construction industry, including floors, walls and ceilings
Electronics
Mechanical engineering and equipment construction
Automotive industry, aviation industry

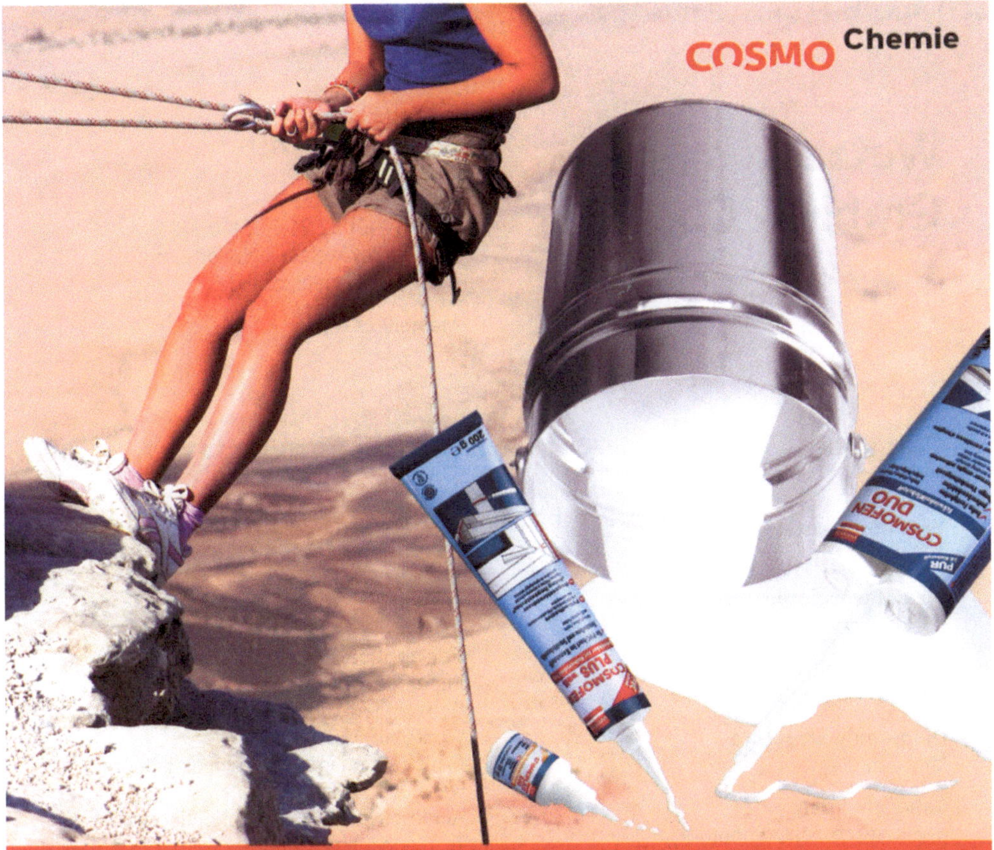

Wetzel GmbH & Co. KG

Hohe Warth 15 – 21
D-32052 Herford
Phone +49 (0) 52 21-7 70 10
Fax +49 (0) 52 21-7 15 46
Email: info@wetzel-klebstoffe.de
www.wetzel-klebstoffe.de

Member of IVK

Company

Year of formation
1935

Size of workforce
65

Managing partners
Michael Kriesten

Ownership structure
family owned

Sales channels
own sales force, distributors

Contact partners
Management:
Michael Kriesten

Application technology and sales:
Yousuf Ali
Klaus-Peter Schneider

Range of Products

Types of adhesives
Hot melt adhesives
Dispersion adhesives
Pressure-sensitive adhesives

For applications in the field of
Paper/packaging
Bookbinding/graphic design
Wood/furniture industry
Construction industry, including floors,
walls and ceilings
Automotive industry, aviation industry
Textile industry
Hygiene

Worlée-Chemie GmbH

Grusonstraße 22
D-22113 Hamburg
Phone +49 (0) 40-7 33 33-0
Fax +49 (0) 40-7 33 33-11 70
Email: service@worlee.de
www.worlee.com

Member of IVK

Company

Year of formation
1851

Size of workforce
230

Managing partners
Dr. Albrecht von Eben-Worlée
Reinhold von Eben-Worlée
Ulrich von Hacht

Ownership structure
private family owned

Subsidiaries
Netherlands, United Kingdom, India, Italy,
Switzerland

Contact partners
Management:
Klaus D. Koehler

Application technology and sales:
Marc Henri Deransart

Range of Products

Raw materials
Additives
Resins
Polymers

Polyurethan Schaumsysteme
Klebstofftechnik

Zelu Chemie GmbH
Robert-Bosch-Straße 8
D-71711 Murr
Phone +49 (0) 0 71 44-82 57-0
Fax +49 (0) 0 71 44-82 57-30
Email: info@zelu.de
www.zelu.de

Member of IVK

Company

Year of formation
1889

Size of workforce
~ 25 employees

Ownership structure
GmbH

Sales channels
Technical Sales

Contact partners
Management:
Dr. Erkut Ilter

Application technology and sales:
Dr. Rüdiger Nowack
Phone +49 (0) 0 71 44-82 57-23
Dr. Natalia Fedicheva
Phone +49 (0) 0 71 44-82 57-39

Range of Products

Types of adhesives
Hot melt adhesives
Reactive adhesives
Solvent-based adhesives
Dispersion adhesives
Pressure-sensitive adhesives

Types of sealants
PUR sealants

For applications in the field of
Paper/packaging
Wood/furniture industry
Construction industry, including floors,
walls and ceilings
Electronics
Mechanical engineering and equipment
construction
Automotive industry, aviation industry
Textile industry
Household, recreation and office

Equipment and
Plant Manufacturing

Bühnen GmbH & Co. KG
Hinterm Sielhof 25
D-28277 Bremen
Phone +49 (0) 4 21-51 20-1 25
Fax +49(0) 4 21-51 20-2 60
Email: info@buehnen.de
www.buehnen.de

Member of IVK

Company

Year of formation
1922

Size of workforce
70

Ownership structure
Private ownership

Subsidiaries
PAM, Incorp., USA
Bühnen, Polska Sp. z o.o.

Contact person
Managing Director:
Hanno Pünjer

Sales Director:
Germany, Austria, Switzerland
Hans-Gerhard Hartje

Sales Director International:
Dr. Valentino Di Candido

Distribution channels
Direct sales, distributor

Range of Products

Application Technology
Hot melt tank applicator systems with
piston pump or gear pump, PUR- and POR-
hot melt tank systems, PUR- and POR-bulk
unloader, hand guns for spray and bead line
application, roller coater, application heads
for beadline, surface, spray, dot, spiral
application and special application heads
with individual customer requirements,
hand-operated glue applicators, PUR- and
POR glue applicators, comprehensive range
of application accessories, customer-oriented
application, solutions.

Hot melt adhesives
The product range includes a
variety of different hot melt adhesive
for almost every application.
Available bases:
EVA, PO, POR, PA, PSA, PUR, Acrylate.
Available shapes:
slugs, sticks, granules, pillows, blocks,
cartridges, barrels, drums.

Applications areas
Automotive, Packaging, Display Manufac-
turing, Electronic Industry, Filter Industry,
Shoe Industry, Foamplastic and Textile
Industry, Case Industry, Construction
Industry, Florists, Wood, Processing and
Furniture Industry.

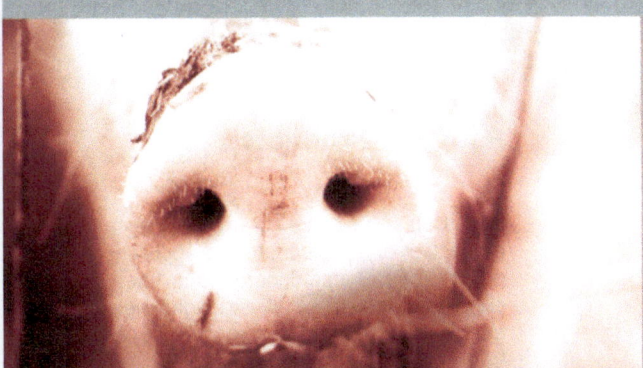

Inatec GmbH

Schneiderstraße 73
D-40764 Langenfeld
Phone +49 (0) 21 73-2 75 75-0
Fax +49 (0) 21 73-2 75 75-25
Email: info@inatec-gmbh.de
www.inatec-gmbh.de

Company

Year of formation
1992

Size of workforce
32

Managing partners
Dipl.-Ing. H. Wallner
J. Rothen
M. Rothen

Ownership structure
1/3 each

Contact partners
Design:
Dipl.-Ing. H. Wallner

Application technology and sales:
Josef Rothen
Michael Brune
Christian Methfessel
Erich Schmidt
Guido Emmerlich

Purchasing:
M. Rothen

Further information
Pilot plant station for testing and development of innovative application solutions are
available.

Range of Products

INATEC has a huge area of expertise in the
production of equipment for the application
of adhesives. The product range includes
melting equipment for all usual adhesives,
from hot to cold, from gelatine to granulate,
from bags to drums; application heads and
dies for wide, line, spray, rotation, contact
and curtain coating; connecting hoses,
control systems and the innovative know-how
for special solutions. From these components, we configure ideal equipment. Or we
develop completely new technologies and
machinery.
For almost every area of application, we
already have the right solution. If not, we
develop it.

Robatech
Gluing Technology

Robatech AG
Pilatusring 10
CH-5630 Muri AG
Phone (+41) 56 675 77-00
Fax (+41) 56 675 77-01
Email: info@robatech.ch
www.robatech.ch

Member of IVK

Company

Year of formation
1975

Size of workforce
More than 460 employees all over the world

Shareholder
Robatech AG, CH-5630 Muri, Switzerland

Ownership structure
Robatech AG, CH-5630 Muri, Switzerland

Subsidiaries and agencies
Represented in more than 58 countries
worldwide
Germany: Robatech GmbH,
Im Gründchen 2, D-65520 Bad Camberg
Phone +49 (0) 64 34-94 11-0
Fax +49 (0) 64 34-94 11-22

Channel of distribuition
Via Head office, subsidiaries and agencies

Contact partners
Management:
Robatech AG, Switzerland:
Marcel Meyer, Thomas Meyer
Robatech GmbH, Germany:
Eberhard Schlicht, Andreas Schmidt

Application technology and sales:
Robatech AG, Switzerland:
Peter Gyger, Sales Director
Ruedi Brun, Marketing Director
Robatech GmbH, Germany:
Eberhard Schlicht, Managing Director

Further information
Production facility in Germany,
Italy and Switzerland

Range of Products

Product and sales program of the company
- Glue application system with piston pumps and gear pumps for hotmelt and dispersions, inclusive necessary equipment.
- Small hotmelt application systems up to 4 liter tank capacity
- Medium hotmelt application systems from 4 to 30 liter tank capacity
- Big hotmelt application systems from 55 to 160 liter tank capacity
- Hotmelt application systems for PUR-hotmelts from 3 to 30 liter tank capacity
- Drumunloaders from 50 to 200 liter tank capacity
- Application technics: Bead application, Surface coating, Spray application, spiral application
- Electrical timing and electrical metering
- Cold glue application systems: Pressure tanks and pump systems

Robatech offers a solution for several industries
Packaging Industry
Converting Industry
Printing Industry
Hygienic Industry
Textile-converting Industry
Woodworking Industry
Building Supplies Industry
Automotive Industry
Various Industries

Sulzer Mixpac

Sulzer Mixpac
Ruetistraße 7
CH-9469 Haag
Phone +41 (0) 81 772 20 00
Fax +41 (0) 81 772 20 01
Email: mixpac@sulzer.com
www.sulzerchemtech.com

Company

Year of formation
Sulzer Mixpac is a merger of successful, former independent companies with already existing business connections and cooperations. The formal acquisition was realized 2007.

Size of workforce
420 employees worldwide

Distribution
Direct distribution to producers of adhesives
Distribution partners for official trade

Contact partners
Europe:
Sulzer Mixpac AG
Email: mixpac@sulzer.com

China:
Sulzer Mixpac China
Email: mixpac@sulzer.com

Untited Kingdom:
Metix UK Ltd.
Email: sales@metix.co.uk

United States:
Sulzer Mixpac USA Inc.
Email: info@SulzerMixpacUSA.com

Range of Products

Manufacturer and supplier of metering, mixing and dispensing systems for reactive multi component material, offering comprehensive systems for various cartridge based 2-K applications.

Cartridge based 2-K systems with different volumes between 2.5 ml and 1.500 ml, with mixing rations of 1 : 1, 2 : 1, 4 : 1 and 10 : 1
Manual and pneumatic dispensers
Meter Mixing systems with various mixing rations of 1 : 1 to 10 : 1
Full range of mixers for cartridge and dispensing machine applications
Brands: Mixpac™, Quadro™, Statomix™

Scheugenpflug AG
Automation and
Dispensing Technology
Gewerbepark 23
D-93333 Neustadt a. d. Donau
Phone +49 (0) 94 45-95 64-0
Fax +49 (0) 94 45-95 64-40
Email: vertrieb@scheugenpflug.de
www.scheugenpflug.de

Company

Year of formation
1990

Size of workforce
180 (worldwide)

Subsidiaries
Scheugenpflug, Inc. USA
Scheugenpflug Ltd. China

CEO
Erich Scheugenpflug, Johann Gerneth

Contact
Sales Germany:
Dietmar Parthum, Phone +49 (0) 94 45-95 64-0
Sales USA:
Dieter Wingel, Phone +1 770 218 08 35
Sales Italy, Atra Srl:
V. Lenardi, Phone +39 (0) 348 516 4770
Sales Spain. Sinutech, S. A.:
Joan Bonfill, Phone +34 93 859 2757

Further information
With more than 180 employees, a successful product portfolio and a worldwide strong
market share, we have reached a leading
position as supplier of resin mix/metering
systems and automation of dispensing
processes performed in normal atmosphere
pressure or vacuum aided. The key com-
petences are high quality and reliable resin
preparation-, feeding- and dispensing units
at a best price/performance ratio. The automation of the processes before and after
dispensing (as: pre-treatment, pick & place,
screwing, curing, testing, a. s. o.) are part

Range of Products

Metering systems
Piston Pump Metering Head
Piston Pump Metering Head Micro and Nano
Piezo Metering Head
Gear Pump Metering Head

Dispensing units
CNCell
RoboCell
DesktopCell
In-lineCell

Vacuum dispensing systems
VDS D (Desktop – Prototypes)
VDS B (Basic – small series)
VDS U (Universal – fast series)
VDS P (Power – high output)

Automation
In-line production units in every level of
automation
Integration of all processes all-round the
dispensing process

Service
Application Centre/Dispensing Samples
Maintenance/Hotline
Teaching
After Sales

---

of our "one supplier philosophy". Scheugenpflug AG supplies and advises reputable
worldwide acting companies of electronic,
automotive, medical tech and consumer
products industry.

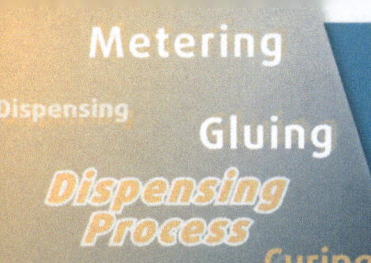
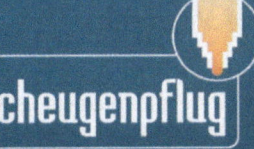

Tigres
Dr. Gerstenberg GmbH

Mühlenstraße 12
D-25462 Rellingen
Phone +49 (0) 41 01-77 78 88
Fax +49 (0) 41 01-77 78 89
Email: tigres@tigres.de
www.tigres.de

Company

Year of formation
1993

Size of workforce
17

Manager
Dr. K. Gerstenberg

Contact partners
Timo Landahl
Email: landahl@tigres.de

Sales
Agencies in Europe and overseas

Range of Products

All technologies of pretreatment using
electrical discharge at atmospheric pressure

Barrier-Discharge (DBD):
"Classical" direct corona,
e. g. for films and foils

Corona-GUN©:
Indirect corona with blown current filaments,
e. g. for sheets (2 m)

Plasma-BLASTER:
Open, potential-free atmospheric plasma,
e. g. glue line, profile, sheets (2 m)

Standard equipment for webs, profiles
and bulk goods

Project planning and construction of
customer adapted solutions

Special developments for R&D and
production

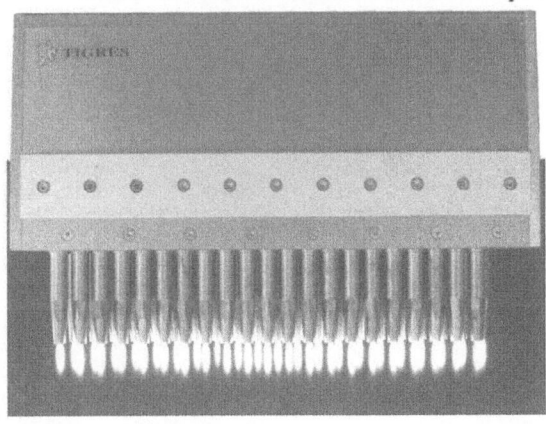

Adhesive Technology
Consultancy Companies

ChemQuest Europe Inc.

Bilker Straße 27
D-40213 Düsseldorf
Phone +49 (0) 211-4 36 93 79
Fax +49 (0) 32 12-1 07 16 75
Mobile: +49 (0)1 71-3 41 38 38
Email: jwegner@chemquest.com
www.chemquest.com

Member of IVK

Company

Contact partners
Dr. Jürgen Wegner,
Managing Director
Email: jwegner@chemquest.com

Dr. Hubertus von Voithenberg,
Managing Director
Email: hvoithenberg@chemquest.com

CV and professional background
under www.chemquest.com

Range of Products

The CemQuest Group is an international consulting firm headquartering in Cincinnati, Ohio, USA, with branch offices in Europe, China, North Africa and Latin America. We specialize in consulting the Adhesives, Sealants, Construction Chemicals and Coatings Industry through all steps within the value chain from raw materials manufacturing through product formulation to all types of industrial and non-industrial end use applications.

Based on in-depth knowledge our service portfolio includes all types of Management Consulting, M&A activities, market research plus market and technology trend analysis. All associates of ChemQuest are experienced professionals from within the Adhesives, Sealants or Coatings Industry, and we are partnering with IFAM Fraunhofer Bremen in certified Adhesives education in North America. For further information and contact details please visit our website www.chemquest.com

Our know-how
– your future!

HINTERWALDNER CONSULTING

Consulting Chemists &
Business Economists since 1956

Hinterwaldner Consulting GbR
Markplatz 9
D-85614 Kirchseeon
Phone +49 (0) 80 91-53 99-0
Fax +49 (0) 80 91-53 99-20
Email: info@HiwaConsul.de
www.HiwaConsul.de

Member of IVK

Company

Year of formation
1956

Managing partners
Dipl.-Chem. Rudolf Hinterwaldner
Dipl.-Kfm. Stephan Hinterwaldner

Ownership structure
Privately held

Contact partners
Management:
Dipl.-Kfm. Stephan Hinterwaldner

Application technology and sales:
Dipl.-Kfm. Stephan Hinterwaldner

Range of Products

Worldwide consultancy and own know-how among others in

- Raw Materials
- Formulations
- Product Developments
- Adhesive Technology
- Adhesives
- Pressure Sensitive Adhesives
- Hotmelt Adhesives
- Glues
- Sealants
- Adhesive Tapes and the like
- Applications
- Polymer Chemistry
- Nanotechnology
- Cosmetics

on the basis of more than 50 years international experiences and own patents as well as technical publications and documentation.

Organizer/Coorganizer:
Munich Adhesives and Finishing Symposium (MKVS) and European Coatings Congress (ECC)

Research and Development

Fraunhofer IFAM

**Fraunhofer Institute for
Manufacturing Technology and
Applied Materials Research
– Adhesive Bonding Technology
and Surfaces –**

Wiener Straße 12
D-28359 Bremen
Phone +49 (0) 4 21-22 46-4 00
Fax +49 (0) 4 21-22 46-4 30
Email: ktinfo@ifam.fraunhofer.de
www.ifam.fraunhofer.de

Member of IVK

Company

Year of formation
1968

Size of workforce
All in all > 380
Department of Adhesive Bonding Techno-
logy and Surfaces: > 200

Ownership structure
The Fraunhofer IFAM is a constituent
entity of the Fraunhofer-Gesellschaft zur
Förderung der angewandten Forschung e. V.

Contact
**Fraunhofer IFAM – Department of
Adhesives Technology and Surfaces**
Acting Director:
Priv.-Doz. Dr. habil. Andreas Hartwig
Deputy Director: Dr. Stefan Dieckhoff

Work Areas

R&D – Contract-research and development
– in all fields of adhesive bonding technology
and surfaces, in addition providing certified
training courses in adhesive bonding technology and fiber composite materials:

Adhesive Bonding Technology:
Dipl.-Ing. Manfred Peschka
Phone +49 (0) 4 21-22 46-5 24
Email: manfred.peschka@ifam.fraunhofer.de

Plasma and Surface Technology:
Dr. Ralph Wilken
Phone +49 (0) 4 21-22 46-4 48
Email: ralph.wilken@ifam.fraunhofer.de

Range of Products

Adhesives and Polymer Chemistry:
Priv.-Doz. Dr. habil. Andreas Hartwig
Phone +49 (0) 4 21-22 46-4 70
Email: andreas.hartwig@ifam.fraunhofer.de

Paint/Lacquer Technology:
Dr. Volkmar Stenzel
Phone +49 (0) 4 21-22 46-4 07
Email: volkmar.stenzel@ifam.fraunhofer.de

Adhesion and Interface Research:
Dr. Stefan Dieckhoff
Phone +49 (0) 4 21-22 46-4 69
Email: stefan.dieckhoff@ifam.fraunhofer.de

Material Science and Mechanical Engineering:
Dr. Markus Brede
Phone +49 (0) 4 21-22 46-4 76
Email: markus.brede@ifam.fraunhofer.de

*Center for Adhesive Bonding Technology/
Training Courses:* Prof. Dr. Andreas Groß
Phone +49 (0) 4 21-22 46-4 37
Email: andreas.gross@ifam.fraunhofer.de
www.bremen-bonding.com

Plastics Competence Center Bremen:
Dr. Silke Mai
Phone +49 (0) 4 21-22 46-6 25
Email: silke.mai@ifam.fraunhofer.de
www.kunststoff-in-bremen.de

*Fraunhofer Project Group Joining and
Assembly FFM:*
Dr. Dirk Niermann
Phone +49 (0) 4 21-22 46-4 39
Email: dirk.niermann@ifam.fraunhofer.de

*Certification Body of the Federal Railway
Authority in accordance with DIN 6701-2:*
Dr. Dirk Niermann
Phone +49 (0) 4 21-22 46-4 39
Email: dirk.niermann@ifam.fraunhofer.de

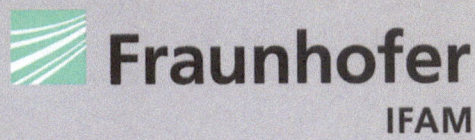
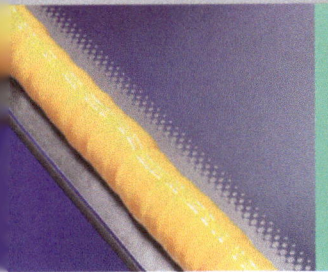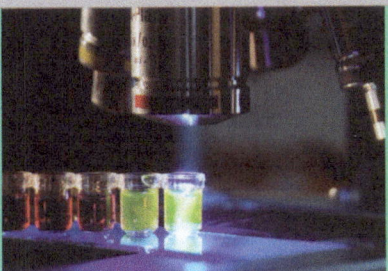

INSTITUTES AND RESEARCH FACILITIES

Research and development

Adhesive bonding technology is a key contributor to the development of innovative products, creating the conditions for new, future-proof markets to open up across all industries. Small Contact:dium-sized companies benefit from new joining methods by developing sophisticated products which give them competitive edge.

In order to profit from the advantages that adhesive bonding technology offers over other methods Contact:, it is important to include the whole process from product planning through quality management to staff training.

This can only be achieved by close cooperation between research institutions and industry so that research results are applied quickly to the development of innovative products and production processes.

The following list includes all known research facilities and institutes which are committed to working with their partners in industry to resolve adhesive issues across a range of areas.

Johann Heinrich von Thünen Institut (vTI)
Federal Research Institute for Rural Areas,
Forestry and Fisheries
Institute of Wood Technology and Wood Biology
Leuschnerstraße 91
D-21031 Hamburg

Contact:
Prof. Dr. Arno Frühwald
Phone +49 (0) 40-7 39 62-6 01
Dr. Martin Ohlmeyer
Phone +49 (0) 40-7 39 62-635
Fax +49 (0) 40-7 39 62-699
Email: martin.ohlmeyer@vti.bund.de
www.vti.bund.de

Deutsches Institut für Bautechnik
(German Institute for Building Technology)
Kolonnenstraße 30 L
D-10829 Berlin

Contact:
Herr Haamann
Phone +49 (0) 30-7 87 30-271
Fax +49 (0) 30-7 87 30-362
Email: jha@dibt.de
www.dibt.de

Fachhochschule Eberswalde
(Eberswalde University of Applied Science)
Wood technology department
Alfred-Möller-Straße 1
D-16225 Eberswalde

Contact:
Prof. Dr. Klaus Dreiner
Phone +49 (0) 33 34-6 55 30
Fax +49 (0) 33 34-6 55 31
Email: kdreiner@fh-eberswalde.de
www.fh-eberswalde.de/holztechnik

Fachhochschule Gelsenkirchen
(Gelsenkirchen University of Applied Science)
Recklinghausen Division
Applied natural sciences department
Laboratory for organic chemistry
and polymers
August-Schmidt-Ring 10
D-45665 Recklinghausen

Contact:
Prof. Dr. Klaus-Uwe Koch
Phone +49 (0) 23 61-9 15-4 56
Fax +49 (0) 23 61-9 15-7 51
Email: klaus-uwe.koch@fh-gelsenkirchen.de
www.fan.re.fh-gelsenkirchen.de

Fachhochschule München
(Munich University of Applied Science)
Process engineering, paper and packaging,
Adhesives and finishing
Lothstraße 34
D-80335 München

Contact:
Prof. Dr. Dirk Burth
Phone +49 (0) 89-12 65-15 58 oder -15 25
Fax +49 (0) 89-12 65-15 02
Email: burth@hm.edu
www.hm.edu

Fogra Forschungsgesellschaft Druck e. V.
(printing research association)
Streitfeldstraße 19
D-81673 München

Contact:
Dr. Eduard Neufeld
Dipl.-Ing. Thomas Kuen
Phone +49 (0) 89-4 31 82-2 71
Fax +49 (0) 89-4 31 82-1 00
Email: kuen@fogra.org
www.fogra.org

FOSTA – Forschungsvereinigung
Stahlanwendung e. V.
(steel application research association)
Stahl-Zentrum,
Sohnstraße 65
D-40237 Düsseldorf

Contact:
Hans-Joachim Wieland
Phone +49 (0) 2 11-67 07-8 56
Fax +49 (0) 2 11-67 07-8 40
Email: hans-joachim.wieland@
stahlforschung.de
www.stahlforschung.de

Fraunhofer-Institut für Fertigungstechnik und
Angewandte Materialforschung – IFAM
(Fraunhofer Institute for Manufacturing Technology
and Applied Materials Research – IFAM)
Wiener Straße 12
D-28359 Bremen

Contact:
Dr. Andreas Hartwig
Phone +49 (0) 04 21-22 46-4 01
Fax +49 (0)4 21-22 46-4 30
Email: helmut.schaefer ifam.fraunhofer.de
www.ifam.fraunhofer.de

Fraunhofer-Institut für Holzforschung –
Wilhelm-Klauditz-Institut (WKI)
(Fraunhofer Institute for Wood Research –
Wilhelm-Klauditz-Institut (WKI))
Bienroder Weg 54 E
D-38108 Braunschweig

Contact:
Prof. Dr. Rainer Marutzky
Phone   (05 31) 21 55-212
Email: info@wki.fraunhofer.de
www.wki.fraunhofer.de

Fraunhofer-Institut für Werkstoff-
und Strahltechnik – IWS
(Fraunhofer Institute for Material and
Beam Technology - IWS)
Klebtechnikum an der TU Dresden-
Institut für Oberflächen- und Fertigungstechnik
Winterbergstraße 28
D-01277 Dresden

Contact:
Dr. Irene Jansen
Phone +49 (0) 3 51-46 33-52 10
Email: irene.jansen@iws.fraunhofer.de
www.iws.fraunhofer.de

Fraunhofer-Institut für
Zerstörungsfreie Prüfverfahren – IZFP
(Fraunhofer Institute for Non-Destructive
Testing - IZFP)
Im Stadtwald, Geb. 37

Contact:
Dr. Bernd Valeske
Phone +49 (0) 6 81-93 02-39 89
Fax +49 (0) 06 81-93 02-59 04
Email: bernd.valeske@izfp.fraunhofer.de
www.izfp.fhg.de

IdM-Institut für dynamische Materialprüfung
an der Universität Ulm
(IdM-Institute für Dynamic Materials Testing
Associated with the University of Ulm)
Helmholtzstraße 20
D-89081 Ulm

Contact:
Prof. Dr. Wolfgang Pechhold
Phone +49 (0) 7 31-50-2 39 90
Phone +49 (0) 7 31-50-2 39 92
Email. wolfgang.pechhold@uni-ulm.de
www.uni-ulm.de/institute/idm

ift Rosenheim
Institut für Fenstertechnik e.V.
(Institute for window technology)
Theodor-Gietl-Straße 7 - 9
D-83026 Rosenheim

Contact:
Dipl.-Ing. (FH) Ulrich Sieberath
Phone +49 (0) 80 31-2 61-0
Fax +49 (0) 80 31-2 61-2 90
Email: info@ift-rosenheim.de
www.ift-rosenheim.de

ihd – Institut für Holztechnologie
Dresden gGmbH
(Institute for wood technology)
Zellescher Weg 24
D-01217 Dresden

Contact:
Dr. rer. nat. Steffen Tobisch
Phone +49 (0) 3 51-46 62-0
Mobile +49 (0) 1 62-2 69 63 30
Email: info@ihd-dresden.de
www.ihd-dresden.de

IFF GmbH
Industrieanlagen, Fügetechnik, Fertigungstechnik
(Industrial plants, joining technology,
production technology)
Krausstraße 22a
D-85737 Ismaning

Contact:
Prof. Dr.-Ing. Christian Lammel
Phone +49 (0) 89-96 99 89-0
Fax +49 (0) 89-96 99 89-29
Email: info@iff-gmbh.de
www.iff-gmbh.de

Leibniz-Institut für Polymerforschung Dresden e.V.
(Leibniz Institute of Polymer Research Dresden)
Büsgenweg 4
D-37077 Göttingen

Contact:
Prof. Dr. H. Militz
Phone +49 (0) 5 51-39 35 41 oder
+49 (0) 5 51-39 35 58
Fax +49 (0) 5 51-39 96 46
Email: bjunge1@gwdg.de
www.wood.uni-goettingen.de

iwb-Anwenderzentrum Augsburg
(Augsburg application centre)
Beim Glaspalast 5
D-86153 Augsburg

Contact:
Dipl.-Ing. Johannes Schilp
Phone   +49 (0) 8 21-5 68 83-20
Fax +49 (0) 8 21-5 68 83-50
Email: info@iwb-augsburg.de
www.iwb.tum.de

Leibniz-Institut für Polymerforschung Dresden e.V.
(Leibniz Institute of Polymer Research Dresden)
Hohe Staße 6
D-01069 Dresden

Contact:
Dr. Rüdiger Häßler
Phone +49 (0) 3 51-46 58-316
Fax +49 (0) 3 51-46 58-284
Email: rhaesz@ipfdd.de
www.ipfdd.de

NMI Naturwissenschaftliches und
medizinisches Institut
(NMI Natural and Medical Sciences Institute)
Markwiesenstraße 55
D-72770 Reutlingen

Contact:
Dr. Bernhard Schröder
Phone +49 (0) 71 21-5 15 30-52
Fax +49 (0) 71 21-5 15 30-62
Email: schroeder@nmi.de
www.nmi.de/de/index.php

ofi Österreichisches Forschungsinstitut für
Chemie und Technik Institut für Klebetechnik
(ofi Austrian Research Institute for Chemistry and
Technology Institute for Adhesive Technology)
Franz-Grill-Straße 5, Arsenal Objekt 213
D-1030 Wien

Contact:
Dr. Werner Preusser
Phone +43 1 798 16 01-780
Fax +43 1 798 16 01-700
Email: werner.preusser@ofi.at
www.ofi.co.at

Papiertechnische Stiftung PTS
(paper technology foundation)
Heßstraße 134
D-80797 München

Contact:
Dr. Renke Wilken
Phone +49 (0) 89-1 21 46-0
Fax +49 (0) 89-1 21 46-36
Email: renke.wilken@ptspaper.de
www.ptspaper.de

Prüf- und Forschungsinstitut Pirmasens e.V.
(Test and Research Institute Pirmasens)
Marie-Curie-Straße 19
D-66953 Pirmasens

Contact:
Dr. Gerhard Nickolaus
Phone +49 (0) 63 31-24 90 00
Fax +49 (0) 63 31-24 90 60
Email: info@pfi-pirmasens.de
www.pfi-pirmasens.de/typo3/index.php

Technische Universität Braunschweig
Institut für Füge- und Schweißtechnik
(Technical University Braunschweig
Institute of Joining and Welding Technique
Braunschweig)
Langer Kamp 8
D-38106 Braunschweig

Contact:
Prof. Dr.-Ing. Klaus Dilger
Phone +49 (0) 5 31-3 91-78 21
Fax +49 (0) 5 31-3 91-58 34
Email: k.dilger@tu-bs.de
www.ifs.ing.tu-bs.de/start

Technologie-Centrum Kleben
TC-Kleben GmbH
(Adhesive technology centre)
Carlstraße 50
D-52531 Übach-Palenberg

Contact:
Dipl.-Ing. Julian Band
Phone +49 (0) 24 51-97 12 00
Fax +49 (0) 24 51-97 12 10
Email: info@tc-kleben.de
www.tc-kleben.de

Technische Universität Berlin
Fügetechnik und Beschichtungstechnik
im Institut für Werkzeugmaschinen
und Fabrikbetrieb
(Technical University Berlin
Joining and coating technology at the Institut
für Werkzeugmaschinen und Fabrikbetrieb)
Pascalstraße 13
D-10587 Berlin

Contact:
Prof. Dr. Lutz Dorn
Phone +49 (0) 30-3 14-2 33 64
Fax +49 (0) 30-3 14-2 31 21
Email: lutz.dorn@tu-berlin.de
www2.tu-berlin.de/fak5/fuegetechnik

TU Dresden
Professur Fügetechnik und Montage
(Professorship for Joining Technique
and Installation)
George-Bähr-Straße 3c
D-01069 Dresden

Contact:
Prof. Dr.-Ing. habil. Uwe Füssel
Phone +49 (0) 3 51-4 63-3 76 15
Fax +49 (0) 3 51-4 63-3 72 49
Email: ftm@mciron.mw.tu-dresden.de
tu-dresden.de/die_tu_dresden/
fakultaeten/fakultaet_maschinenwesen/
iof/fue

TU Kaiserslautern AWOK Arbeitsgruppe
Werkstoff- und Oberflächentechnik Kaiserslautern
(Work Group for Materials and Surface
Technology (AWOK) at the Technical University
of Kaiserslautern)
Gottlieb-Daimler-Straße
Geb. 44, Raum 334
www.mv.uni-kl.de/awok
D-67663 Kaiserslautern

Contact:
Prof. Dr.-Ing. Paul L. Geiß
Phone +49 (0) 6 31-2 05 41 17
Fax +49 (0) 6 31-2 05 39 08
Email: Geiss@mv.uni-kl.de
www.mv.uni-kl.de/awok

Universität Kassel
Institut für Werkstofftechnik, Kunststofftechnik
(Institute for Materials Engineering,
Plastics Technology)
Mönchbergstraße 3
D-34125 Kassel

Contact:
Prof. Dr.-Ing. Hans-Peter Heim
Phone +49 (0) 5 61-8 04-36 70
Fax +49 (0) 5 61-8 04-36 72
Email: heim@uni-kassel.de
www.uni-kassel.de/fb15/ifw/reil/
welcome.ghk

Universität Paderborn
Laboratorium für Werkstoff- und Fügetechnik
(Laboratory for materials engineering and
joining technology)
Pohlweg 47 - 49
D-33098 Paderborn

Contact:
Dipl.-Wirt.-Ing. Dominik Teutenberg
Phone +49 (0) 52 51-60-30 31
Fax +49 (0) 52 51-60-32 39
Email: sekretariat@lwf.upb.de
www.lwf.uni-paderborn.de

Wehrwissenschaftliches Institut für
Werk-, Explosiv- und Betriebsstoffe
(Bundeswehr Research Institute for Materials,
Explosives, Fuels and Lubricants (WIWEB))
Institutsweg 1
D-85435 Erding

Contact:
Dr. Jürgen von Czarnecki
Phone +49 (0) 81 22-95 90-0
Fax +49 (0) 81 22-95 90-39 02
Email: juergenvonczarnecki@
bundeswehr.org
www.bwb.org/wiweb

ANNUAL REPORT 2008

German
Adhesives
Association
Industrieverband Klebstoffe e. V.

Market report

Between the year 2000 and 2007 the adhesives and sealants demand in Germany grew in average 4 - 5 % p. a., and the export of adhesives "Made in Germany" was a powerful driver for the German adhesives industry. Despite the strong Euro, exports in 2006 and 2007 saw a year-to-year growth of 9 percent, reaching a total of almost 1 billion Euros.

A clear shift in main export market from Eastern Europe to Asia could be discerned. This boom in export not only safeguards the economic basis of the German adhesives industry, but it also underlines the market and global technology leadership of this innovative German adhesives and sealants industry.In spite of the global financial and economic crisis which reached the German adhesives and sealants industry during the second half of 2008 the industry could again increase the production volume by 2 %, compared to the previous year.

During the second half of 2008 and the first quarter 2009 the global fiscal and economic crisis affected mainly the adhesive demand in the market sectors transportation and electronic (approx. - 25 % and -15 %), while the demand in the other key market sectors remained fairly stabilized. The adhesive demand in the segments "wood & furniture", „DIY", as well as in "paper & packaging" stand to benefit from the so called "cocooning effect", and in the market segment "building/construction" the German adhesives and sealants industry benefits from the economic stimulus package introduced by the German government, aiming at refurbishing of public buildings such as schools, universities, etc.

In total the German industry is confident not to loose more than 2-3 % of its turnover in 2009. The industry's total sales in Germany (adhesives, sealants, tapes and cement based systems) still accounts for almost 3 billion Euros, generating an added economic value of more than 325 billion Euros.

From a European perspective, more than 25 % of all adhesives produced in Western Europe are consumed in Germany. In turn, the German adhesives industry holds a 10 % market share of the global adhesives market. Apart from the fact that the economic key indicators already draw a positive picture for the fourth quarter 2009 there are numerous opportunities for innovations in adhesive bonding which have been developed between the adhesives industry and its customers during the past "quiet" months.

The electronics industry is one of the markets interested in innovation: for instance, to produce the new biometric passports, special adhesives were developed which help glue microchips with biometric data (personal data and digital face characteristics) directly onto the chip housing and contact the housing electronically with wires.

The capture, storage and recall of personal and product data on travel and transport documents, event tickets or warehouse items is gaining increasingly importance. The launch of the flip chip technology and electrically conductive systems which have been especially developed to glue microchips and smart labels have enabled series production of this technology and as a consequence the development of new applications.

This is another positive signal to the German adhesives industry to continue developing new markets on an international level using the bonding technology and to pursue product innovation and engineering.

Committee work

General Assembly

The Annual General Meetings in Bamberg and in Cologne were marked by the positive economic situation and robust market development. Even the increase in German value added tax to 19 % could not stop the economic upswing. While companies were able to increase competitiveness through structural changes, the growth in private consumption was somewhat slower. There is some uncertainty as to whether exports will help support economy as usual, and whether private consumption will steadily increase.

FEICA's move to Brussels and the separation of offices has led inevitably to changes in staff. The new FEICA management will still depend on the support of the German association in European matters. It is too soon to say whether this decentralisation will be compensated by the benefits of having a Brussels-based office.

Despite a few mergers, the number of members has continued to increase. This emphasizes the association's position as a well-known and respected representative for the industry and supports continuity. The German adhesives industry in the last few decades has seen robust economic development, and has shown itself to be a broad, innovative industry, the location of which offers a range of benefits: Germany is a large domestic market with numerous key customers, leading machine manufacturers and raw material producers as well as leading research and education institutions.

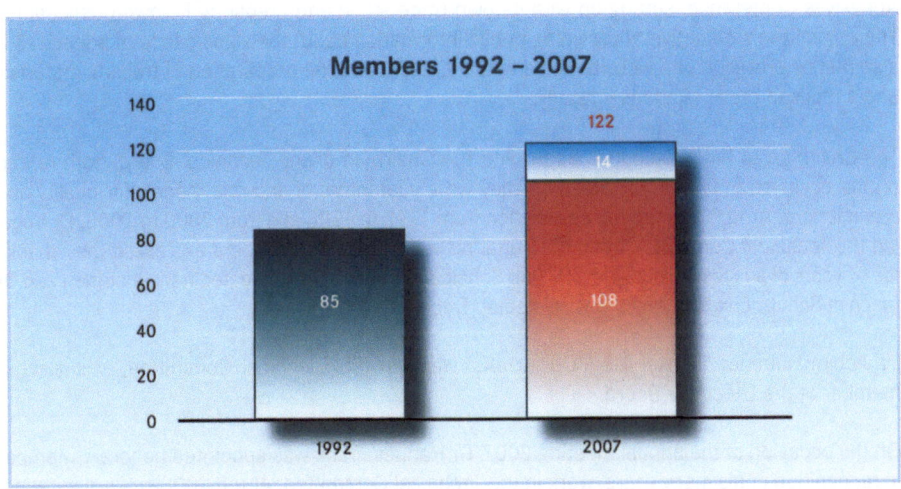

Members 1992 – 2007

The European chemicals regulation REACh which has been in force since June 2007, together with a number of chemistry associations, helped achieve some fairly positive results. Now the emphasis is on strengthening communication with raw material producers and customers. To this end, guidelines and leaflets have been and are being developed, and a helpdesk has been organized in conjunction with the German Chemical Industry Federation (VCI) to provide individual consultancy to companies. The high adaptability and outstanding competence of IVK is demonstrated by focussing on key themes and its positioning as a proficient international resource, for instance giving five lectures about "REACh" at WAC in the USA and by participating in American web seminars.

Through the IVK, the German adhesives industry embraces its role as leader by also pursuing industry-wide objectives. For instance, information systems and codes like GISCODE have been established in order to help protect employees, consumers and the environment. Other key examples of responsibility demonstrated by the adhesives industry include the pursuance of sustainable substitutes for solvent-based products, the establishment of the EMICODE identification system and the development of products free from APEO/DIBP.

European legislation also seeks to find substitutes for solvent-based products, as can be seen by the plan to extend the Decopaint directive to cover specific adhesive segments, in addition to the regulations already in existence. With this initiative, the EU Commission is addressing mainly adhesives for floor coverings and parquet floorings. The German adhesives industry is in a position to pick up the gauntlet and to bring their innovative new technologies to the market.

Public relations are another focus of IVK's work, and the organization constantly develops and improves leaflets about such topics as food law, good manufacturing practice, construction adhesives and the handling of so-called "superglue". They also produce "Klebstoff Nachrichten" (Adhesive News) and organize the participation of the German adhesives Industry at the World Adhesives Conference, setting up with its own trade stand and preparing its own publications. The committee's objective of providing public information about the significance of adhesives is also met by provision of a website and a virtual press room, the preparation of training materials and a focused press relations strategy.

The Chairman of IVK's Executive Board and the Chairman of the Technical Board, both retired on the occasion of the annual meeting. They were guarantors of and role models for quality and commitment. Arnd Picker led the association from 1992 to 1998 and from 2000 to 2008, Dr Vogel led the Technical Committee from 1994 until his retirement. All members expressed their thanks for 14 years of successful activities. Dr Ralf Schelbach, Jowat AG was elected new Chairman and Dr Joachim Schulz, Eukalin GmbH, was re-elected Deputy Chair.

Dr Achim Hübener, Henkel AG & Co. KGaA, Head of the Technical Committee, was elected member of the Executive Board.

On the occasion of the annual meeting 2007, Dr Hannes Frank was appointed honorary member in recognition of his many years' work in the Technical Committee. At the 2008 general meeting,

Dr Rainer Vogel also became a honorary member as a result of the commitment he had shown to the association.

The annual meeting decided that membership fees, which had remained unchanged since the inception of the association, should also be maintained for 2008.

Executive Board

The Executive Board of the German Adhesives Association is involved in an assessment of and discussion by experts about the economic and technological trends in the key market segments of the adhesives industry. It also monitors and analyzes the activities undertaken by the different commercial and technical committees of the association.

The Executive Board aims to always respond quickly to new economic and technological conditions by adapting the organization structure of the association and its committees, seeking to ensure that the organization is always as efficient as possible.

Through a focused implementation of the positioning concept "Qualified Market Expansion", the Executive Board has managed to position the association as a competent platform in industrial expertise for scientific institutions, system partners, education organizations/trade shows, and congresses. Regular dialogue with all relevant system partners and the scientific community has become an integral part of the association's profile. What is more, the training provided by Fraunhofer Institute for Manufacturing Technology and Applied Materials Research IFAM, Bremen, which is strongly supported by the Adhesives Association, has become a popular and renowned programme even beyond the borders of Germany. Companies using adhesives see an increasing need for their staff to become qualified in the use of technically sophisticated adhesives, and thus the adhesives industry benefits from highly-skilled system partners. To date, the Fraunhofer scientists have trained more than 2,500 adhesive practitioners, experts and engineers.

The project to make adhesives training international, which was initiated jointly with the Fraunhofer-Gesellschaft, was successfully completed in 2007. The course materials at various levels (practitioner, expert, engineer) have been translated into English and Chinese and adapted to match the various international standards. In addition, native speakers have been trained to become trainers. Following the launch of the first international pilot classes in the USA and South Africa, regular specialist adhesives training courses will be held from the second quarter of 2008 onwards in the USA and in China in cooperation with qualified partner organizations.

By developing this training standard, which can be applied right around the world and implementing a suitable training system on a global scale, the IVK's Executive Board has once again demonstrated that the German adhesives industry holds a key position on the international scene.

International cooperation can also be seen by the participation of the Industrial Association and its members in major international conferences. Under the coordination of the Executive Board members and staff of the association supported the World Adhesives Conference held in Miami, Florida, USA in 2008 by giving a number of well-attended speeches, thereby amply demonstrating the technological leadership of the German adhesives industry in this international setting. At an exhibition accompanying the conference, the Industrial Adhesives Association, the trade magazine "Adhäsion Kleben & Dichten", Fraunhofer Institute IFAM and GEV jointly introduced themselves to an international expert audience as a global competence platform using the slogan "Made in Germany".

Over the course of a number of discussions involving experts from international adhesives associations from the USA, South America and Asia, it has become very clear that the German adhesives industry is not only perceived as global technology leader but is also highly respected as a competence platform for "Responsible Care®". Long before legal regulations came into force, the German adhesives industry developed, with the agreement of its partners, practical concepts to adequately protect the environment, workers and consumers and successfully implemented these measures in line with the strategic guidelines. The German adhesives industry meets its responsibilities towards the environment and customers along the value chain with initiatives like EMICODE and GISCODE and also by separating solvent consumption from adhesives production. As a result of these kinds of schemes it has come to hold a leading role worldwide. To the Executive Board of the German Adhesives Association, getting the balance right between technological leadership and social competence is a major key to successful and credible positioning of the adhesions industry. This position ensures that the association has access at all times to reliable and important information, for instance about future legislative projects. In addition, the members of the association have the opportunity to open up new markets both within Europe and worldwide by providing products meeting the long-term goal of protecting the environment, workers and consumers.

Following successful association with the Austrian Adhesives Association (VÖK), the Board of Directors has now entered into intense discussions with the Swiss Adhesives Association, with the aim of formalizing the good cooperation which these associations have maintained for a number of years. These associations are part of the strategy proposed by the Executive Board to open the German Adhesives Association to European adhesives manufacturers, and as a result, adhesives manufacturers from Britain, the Netherlands and Turkey have affiliated with the German Adhesives Association over the past few months.

The successful development of GEV (Association for the Control of Emissions in Products for Flooring Installation) is another major milestone in increasing the profile of the German Adhesives Association. The EMICODE system and related testing has meanwhile become accepted in other European countries and is gaining increasing approval internationally. Given the current context of ongoing discussions at a European level dealing with "indoor air", GEV provides an excellent technical foundation for the adhesives industry.

As the Executive Board proposed, GEV is now open to adhesive systems from market segments other than fields related to construction. For this reason, it has changed its name to GEV = Association for the Control of Emissions in Products for Flooring Installation, Adhesives and

Building products). Given this new, broader role, GEV has already developed and published criteria for measuring and assessing VOC emissions from sealants. Recently the experts at GEV have issued test criteria and requirements for parquet sealing agents.

The ongoing emphasis the German Adhesives Association places on public relations is impressively reflected in the half-yearly IVK press reviews. The documented publications quoted in them show clearly that the Public Relations Committee has managed to sustainably arouse the interest of editors and journalists in the trade and the consumer press in „adhesive" topics and to motivate them to report on this fascinating bonding technology and the variety of applications in all areas of day-to-day life.

In addition to producing the DVD "Faszination Kleben", a printed newsletter has been launched, called "Kleben fürs Leben" *(Stuck for Life)*, the third edition of which was issued in late summer 2008. Both the film and the newsletter strictly follow the strategy defined by the Executive Board: They describe the variety of adhesive applications and the huge innovation potential of the German adhesives industry, but are always neutral in terms of companies and products. Complimentary copies of the DVD and newspaper articles are available to association members on request.

The Executive Board feel that the constantly increasing number of delegates at the events organized by the German Adhesives Association and the admission of 15 new member companies over the past 16 months indicate that the German Adhesives Association is well positioned and is seen as providing profitable input and a useful resource for the adhesives industry.

Technical Board (TB)

At a total of six meetings held in 2007/2008, the Technical Board (TB) discussed in detail the work of the technical commissions, subcommittees and ad-hoc bodies. Furthermore, the association made proactive contributions to various interdisciplinary subjects after an intense discussion in the Technical Committee.

The work of the Technical Board was focused around activities dealing with the new European Community Regulation on chemicals "REACh" (Registration, Evaluation, Authorisation and Restriction of Chemical substances). After the EU commission issued its white book describing the basics of its new chemical policy, and, after lengthy discussions, the REACh Regulation 1907/2006 was publicized on 18 December 2006 in the Official Journal of the European Union and entered into force on 1 June 2007. A central item of this draft is that all members of the value chain are responsible for the safe use of their product within the framework of intended use. Thereby the tasks and objectives were clearly shifted towards downstream users. Until the end of 2006, the Technical Board was heavily involved in the process of forming political opinion through close cooperation with FEICA, other German and European industrial associations and the Downstream User of Chemicals Coordination Group, DUCC. From 2007 onwards, TB focused its main activities within the framework of this cooperation on shaping the REACH

Implementation Projects (RIP) to be established by the Commission. Within the framework of RIP 3.2, a system was developed in cooperation with FEICA to determine and demonstrate for thousands of adhesive preparations, each containing a number of substances, the conditions for safe use of preparations. In order to allow this method to be effectively used, it must be mapped into a software system by an external provider. Having finalized the most important implementation guidelines for the industry, the Board's work is now mainly focused on helping members implement the regulation. Current major IVK activities within the framework of REACh involve developing generic exposure scenarios for adhesives, with the aim of making the system much simpler for adhesives manufacturers: instead of preparing a separate exposure scenario for each of their products, they will be able to select from amongst a few comprehensive adhesive-specific exposure scenarios which have been developed by the association. These rather abstract, so-called "generic exposure scenarios" will help in communicating with raw materials manufacturers. In addition, industry-specific formats for an extended material safety data sheet are being developed to allow customer-specific communication with adhesives processors.

IVK members receive information about REACh through the REACh portal and the service hotline of the German Chemical Industry Federation (VCI). They also receive the IVK newsletter dealing with adhesive-specific topics and are invited to attend seminars which IVK organizes in cooperation with training providers.

The EU Commission is set to introduce the new globally harmonized system for classifying and identifying chemicals (GHS) at the same time as REACh (CLP-VO). The launch of this system involves major structural changes compared with today's EU system. Among other things, this entails new risks, new classification methods (e. g., "bridging"), new classification criteria, new identification elements and a new (modified) material safety data sheet. GHS has clear benefits, but formal reasons alone (due to new or revised criteria) mean that there will be more hazardous substances and preparations). Many products will have to be relabelled and two systems will exist in parallel for a longer transition period. For adhesives manufacturers, the simultaneous introduction of GHS and REACh will result in two key issues. First, a great deal of transition work will be entailed for preparations. Data loopholes for single components of preparations lead to very stringent GHS classification if these values are calculated automatically. Expertise is therefore necessary in order to classify preparations correctly in the GHS system. REACh will provide a range of new data throughout the 11 year introduction period. Each preparation will have to be reclassified several times under REACh as new data is generated. What is more, many European and national regulations make reference to classification and identification, for instance the German statutory order on hazardous incidents, the chemicals prohibition ordinance, TA-Luft (5.2.5), water protection and household regulations, various TRGS (e. g., warehouse regulations) as well as a range of EU regulations and directives. If these rules and standards are not adapted, there may be serious consequences for approval procedures, construction activities, organizational activities, etc. Furthermore, regulations which have specifications related to GHS must be reviewed and updated prior to the introduction of GHS. With the interests of the German and European adhesives industry at the forefront, IVK is actively involved in shaping the implementation guidelines, namely within the framework of VCI and, on a European level, by working with FEICA and DUCC (Downstream User of Chemicals Coordination Group).

VCI, together with other professional associations, was requested to investigate options to reduce VOCs in an attempt to diminish the risk of summer smog, as part of an initiative by the Federal Ministry for the Environment, Nature Conservation and Nuclear Safety. The Technical Board looked at this topic in depth and was able to show, using the data from a survey of solvent consumers, that the aim of reducing VOCs by 70 % by 2007 (based on 1988 consumption) had been achieved much earlier. The solvent statistics are compiled on a biannual basis and are referred to in discussions with national and European authorities and institutions, for instance when new legislative projects are introduced.

The EU project MIGRESIVES II, which was launched in February 2007, deals with the question of whether and to what extent the components of adhesives used to produce food packaging can migrate into food. The aim of this project from an industry perspective is to develop a reliable and value for money mathematical model which will help to adapt adhesive formulas and applications so as to ensure safe use. With this, expensive measurement should become less necessary. The support conditions mean that the organization, coordination, financing and execution of the project are the sole responsibility of small and medium-sized enterprises (SME) and associations.

IVK, as one of the project partners, however, ensures the participation of raw material producers and multinational corporations within the limits of what is legally possible. The EU project FACET (Flavours, Additives and Contact materials Exposure Task) deals with the development of scientifically founded instruments so as to be able to estimate the exposure of humans to chemicals from food (e. g., what kind of packaging is used for which food, packaging design). As adhesives form part of packaging, IVK supports the Association of European Adhesives and Sealants Manufacturers (FEICA) in their commitment to the FACET project.

With regard to application in food/articles intended to come into contact with foodstuffs, DBIP was removed from the list of additives and monomers for plastics permitted to come into contact with food in 2007 (annex III of directive 2002/72/EC). For all other areas, a classification for products which are toxic to reproduction, category 2, and for identifying preparations containing 25 % DIBP or more with T, R61 will be applicable in future (presumably from 1 June 2009 onwards). The mixed recycling of food and non-food packaging, together with the use of DIBP as a softener for homopolymer dispersions, for instance in adhesives for non-foodstuff packaging, a high DIBP content can today be found in recycled paper. In order to be able to use recycled fibres for food packaging in the future, the use of DIBP should be largely avoided throughout the paper production chain. TB therefore supports a voluntary negotiated agreement by IVK members to stop offering adhesives containing DIBP to the paper and board producing or processing industry from 1 January 2009 onwards.

The Technical Board has looked in depth at the topic of "adhesive training" and strongly recommends that formal training leading to qualifications be provided to increasing numbers of adhesive users and introduced at a European level. In the same vein, the TC supports a project of Fraunhofer Institute IFAM which aims to make the "Adhesive Bonding Engineer" training available internationally.

The Technical Board has also been involved in the following:

- IMDS - International Material Data Sheet (automotive engineeering)
- Standardization initiative "product information"
- "Blauer Engel" – Environment labelling for adhesives
- Approval procedure for construction products
- MDI/MDI identification in consumer products
- Support to various government departments
- Workers and consumer protection
- EU Risk Assessment
- Assistance with a number of projects of the EU commission at national level
- Standardization on national, European and international level (DIN/CEN/ISO)

Technical Committee
Building and Flooring Adhesives (TKB)

The Technical Commission Building and Flooring Adhesives (TKB) of the German Adhesives Association (IVK) represents the interests of manufacturers of building adhesives and dry mortar systems who are members of IVK. The committee liaises with the authorities, craft guilds, employers' liability insurance associations as well as other industrial organizations and standardization committees. The aims are to establish technical standards, to influence regulations involving the chemical law, to participate in developing legislation, to promote technical progress whilst safeguarding users and the environment, to provide technical support and information to our customers in the building trade, and to promote the use of building adhesives and mortar systems through objective technical information.

TKB's activities at a glance:

- Standardization of levelling compounds, tile adhesives, flooring adhesives and parquet flooring adhesives.
- Coordination with German association for construction chemicals on mortar system issues.
- Organizing/attending technical information events for flooring, tile and parquet flooring tradespeople and people from related trades.
- Production of TKB publications on topics related to current applications and building law as well as standardization, environment and user protection.
- Handling a range of application-related queries in the fields of mortar systems and building adhesives as well as related trades such as screed, flooring/parquet.
- Dealing with matters related to building law, for instance general approval of tile, flooring or parquet adhesives by the building inspectorate (CE-labelling)
- Dealing with topics related to workers and environmental protection.

1. TKB department Standardization

TKB members are involved with CEN/TC 67/WG 3 (tile adhesives), through which the European standards for mortar systems are being developed. The committee is also involved in regularly revising major test standards for tile adhesives as well as formulating the important DIN EN 12004 standard, which defines the requirements for tile adhesives. The work of this committee also established DIN EN 14891 for sealing compounds.

1.2. TKB department Coordination with German Association for Construction Chemicals (KKF)

As part of the group of organizations involved in coordinated matters related to tile laying materials (KKF), TKB, together with German Association for Construction Chemicals, is working mainly on a CWFT classification for the fire behaviour of levelling compounds and tile adhesives, which would clearly reduce the number of tests and thereby costs. Furthermore the committee pushed for the grout standard DIN EN 13888 to be made mandatory, and also assessed the relevance of the "Decopaint" directive for primers in flooring technology, finally deciding that it was not relevant to the industry's field of activity.

2. TKB department Events

TKB Conferences

The traditional TKB conferences are held at the Frankfurt Stock Exchange, which are among the most prestigious events of the German flooring industry. An audience of some 200 people listened to presentations on topics related to the application of technology and the protection of workers and the environment. In addition to TKB members, there were a number of presentations from renowned guest speakers, which received good coverage in the trade press.

TKB Round Table Talks on Flooring Technology

The representatives of major trade associations and the trade press as well as some TKB representatives gathered for the TKB round table talks on flooring technology. This informal event was very well received and the number of attendees had to be limited to 25. The talks were quite informative, covering a range of technical topics ranging from screed to flooring, parquet, adhesives and levelling compounds, as well as some trade political matters.

2. TKB department Publications

TKB Leaflets (Technical Briefing Notes)

TKB leaflets on application topics are important reference material for craftspeople and experts. A first draft of a new leaflet on levelling compounds was almost completed, a new leaflet on MDF (medium density fiber board) is currently being developed to replace a previous edition on particle board. Of particular interest is the updated "Bonding of Parquet Flooring" leaflet, which describes the state of the art in detail, including all new parquet standards and the latest developments in adhesives including application recommendations.

2. TKB department Application Issues

Moisture Measurement for Fast Floor Screed

There are no generally applicable instructions or recommended values available for measuring the moisture or assessing the setting of fast floor screed. No solution has yet been found to cover the whole range of products, given the complexity and diversity of systems sold. The manufacturers' recommendations are still the key reference in this area.

Working Committee TKB – FEB – Craft

The discussions with the association of manufacturers of elastic flooring (FEB) and craftspeople which had got off to a fruitful start lost some momentum, as the application departments of the flooring manufacturers displayed a tendency to reduce headcount. At some meetings there was a lack of technically qualified delegates able to jointly tackle the technical issues. Major topics included the dimensional stability of PVC flooring, the exchange of information regarding product modifications and the establishment of a standardized recommendation system.

CEM I/CEM II Cement

The production of cement screed increasingly uses, in addition to pure Portland cement (CEM I), Portland composite cement (CEM II) of various formulas. The slower hydration gives rise to a number of issues involving moisture measurement and setting. The TKB is looking into these issues in order to be able to provide craftspeople with substantiated re-commendations for the foreseeable future.

2. TKB department Standardization

Within the framework of CEN/TC 193/WG 4, TKB members were involved with the standardization of flooring adhesives. They worked on DIN EN 13851 for testing levelling compounds as well as prEN 13415 which defined the conductivity of flooring adhesives. For DIN EN 14293 on "parquet adhesives", a regular revision is being considered.

2. TKB department Building Law

Approval of Building Adhesives by the Building Inspectorate

CEN/TC 351 works on implementing the mandate M/366 of the European Commission to execute requirement no. 3 "Hygiene, health and environment" of the European Construction Products Directive. This threatens to make all flooring materials subject to approval by the building inspectorate. This would imply the publication of formulas, external supervision and assessment of environmental impact pursuant to the AgBB scheme.

Product specific issues are being referred by TC 351 to the expert TCs, in this case TC 193/ WG 4. Here, TKB members are hard at work trying to avert the requirement to obtain approval by the building inspectorate or at least to ensure that the expensive test in line with the AgBB scheme is replaced by the less costly Emicode system which contains all relevant product data.

TKB-AG fireproof Adhesives

Flooring for public areas is subject to approval by the building inspectorate. One criterion is the fire test. In this case, the behaviour of flooring adhesives in the event of fire is being analyzed. It is the compound comprising adhesive and flooring which is submitted to testing, which means that there will be a countless number of combinations to be tested. In order to avoid this, TKB, together with TFI, has established a working committee whose goal is to define and gain acceptance of a class of "fireproof adhesives", which are classified for instance by their peel resistance and their Emicode. The work in this area is still ongoing.

Approval of Parquet, Parquet Surface Treatment Products and Parquet Flooring Adhesives by the Building Inspectorate

Since March 2009, parquet and parquet sealing have been subject to approval by the building inspectorate, in line with DIN EN 14342. This requirement implies meeting the requirements of no. 3 "Hygiene, health and environment" of the European Construction Products Directive. One consequence of its implementation is the threat to make parquet adhesives subject to approval by the building inspectorate. TKB is considering this situation in depth, controversial discussions currently being held with DIBt. Firstly the relevance of DIN EN 14342 for parquet adhesives must be generally questioned. Then, it could be argued that approval by the building inspectorate might only be needed within the framework of fire testing (glued compound). TKB's seeks at least acceptance for the classification of parquet adhesives so as to minimize testing and also to have Emicodes accepted as an alternative to the AgBB scheme. The work in this area is still ongoing.

3. Protection of workers and the environment/3.1. Gisbau

TKB again maintained excellent cooperation with Gisbau (Employer's Liability Insurance Association for the Construction Industry). The Giscode system which was established by TKB and Gisbau is a good, simple, widely-accepted instrument designed to meet the requirement for research in line with the German Ordinance on hazardous substances and subsequent regulations (GefStoffV and TRGS 440). A new Giscode class, S 0.5, applicable to parquet adhesives with reduced solvent content, has meanwhile been established, following some political arguments. Products of this class substitute products with a high solvent content as they comply with workplace exposure limits.

The new RS 10 Giscode group applicable to flooring materials containing methoxysilane was developed so as to allow classification of this product type which is increasingly gaining importance. A toxicological assessment of this product class is currently in progress, as is testing of the emission behaviour. There is reason to hope that the assessment of issues to protect workers will be so positive that SMP products will be named as substitutes for flooring materials with a high solvent content. The broad range of measured data which Gisbau has compiled with respet to product emissions as well as the description of exposure scenarios and protection of workers could be helpful for the adhesives industry in preparing the data required for REACH. TKB will continue to work with Gisbau on this matter.

3. Protection of workers and the environment/3.2. Chromate in Cement Products

The cement industry association, Deutsche Bauchemie *(German Construction Chemicals Association)*, and the association of dry mortar manufacturers have set up a working committee whose aim is to improve the test method for determining chromate in cement and dry mortar. The test method used in DIN EN 196-10 will probably replace TRGS 613. In mortar systems, the various components of a mix have a significant impact on measuring results, shelf life, and the effectiveness of reducing agents. Furthermore, the definition of the cement portion in mixes is interpretable. The different reducing agents have different levels of effectiveness and stability depending on the actual mortar system. Currently producers therefore cannot rely on chromate reduction by the cement industry alone but must rather make adjustments and tests of their own. TKB representatives are participating on the working committee, where they represent TKB's interests.

3. Protection of workers and the environment/3.3. MDI Identification

Threatening restrictions concerning the identification of MDI in 30th ATP including classification as Carc. Cat. 3 and R 40 (suspicion of carcinogenic effect) jeopardizes the existence of a whole class of floor laying materials with outstanding technical properties: "hard elastic" 1 K and 2 K PU parquet and flooring adhesives. By modifying the way they are identified, the Emicode classification of these products would no longer apply. TKB has started to compare typical exposure scenarios during the use of such adhesives for flooring with marginal conditions which have led to restrictions in terms of identification. For this purpose, Gisbau's insight into worker protection for these products, which can be seen to be harmless in the context of exposure by flooring technicians, is helpful. The work has only just started and seeks to show whether it is legally and morally possible to keep this product group within the range of flooring technology. In addition, work is underway on developing technical solutions so as to circumvent identification by using clean prepolymers.

4. GEV

As GEV is basically rooted in TKB and many TKB members are also affiliated with the Technical Council of GEV, we wish to briefly present GEV's activities.

The classification criteria were revised, and the EC1 R class introduced. Major topics included the execution of requirement no. 3 "Hygiene, health and environment" of the European Construction Products Directive as well as wrestling with DIBt regarding the design of the AgBB scheme and the acceptance of Emicode as an equivalent alternative, as has been described previously.

The various European activities dealing with indoor air were critically monitored in order to avoid, if at all possible, the imposition of additional requirements for floor laying materials. The Emicode system was extended to cover sealants. While it was originally established for flooring adhesives, it now comprises the following product groups: primers, levelling compounds, tile adhesives, joint grouting, joint sealants, flooring and parquet adhesives, underlay, self adhesive tapes and water-based parquet coatings.

An international conference held on the occasion of GEV's 10th anniversary further strengthened the internationalization of the Emicode system. Increased awareness and acceptance of this system means it is likely to become an accepted standard for product assessment with a view to environmental protection.

Within the framework of public relations, a brochure with 30 Q&As on Emicode was revised.

Technical Committee
Committee "DIY and Consumer" Adhesives (TKHHB)

The Technical Committee "DIY and Consumer" Adhesives meets on a regular basis with the respective marketing group and monitors a range of legislation activities at a European and a national level. The commission is focused on regulations and topics concerning adhesives in packaging as far as these are intended for private consumers. Major topics include:

- The identification and packaging of adhesives and the obligation to provide information (chemicals act)
- The limitations with regard to the marketing and use of adhesives
- The requirements in certain areas where adhesives are applied (equipment product safety act/toys regulation/medical products act)
- Other normative/legal requirements on adhesives

The identification and packaging of adhesives and the obligation to provide information (chemicals act)

According to § 13 – 15 a of the chemicals act, the manufacturer or importer must
- Classify substances and preparations prior to marketing them
- Package them according to the classification and
- Label them.

The technical regulation for hazardous materials, TRGS 200, sums up the applicable regulations:
- Special identification regulations for substances and preparations which are available to everyone, 6.7
- Alleviation and exemption from identification, 7.1
- Carrying out identification, 9

In accordance with § 16 e of the chemicals act, intoxication information and treatment centres must be notified. Manufacturers or importers of certain preparations which must be labelled according to § 3 a clause 1 ChemG must inform the Federal Institute of Risk Assessment of
- The trade name,
- The composition,
- The label,
- The use,
- Recommendations on precautionary measures during use and first aid in the event of accidents.

Furthermore, information must be provided which might be of importance in treating conditions which might result from contact with a preparation. Later changes must also be reported. The notification requirement for cleaners and solvents is even more stringent.

Requirements for certain Areas in which Adhesives are applied -
(Medical Products act, Equipment Product Safety act/Toys Regulation)
The general rule is: A product may only be marketed when it is so designed that in the event of the intended use or foreseeable incorrect use, the health and safety of users or third parties is not at risk.

This applies to manufacturers of articles ready for use; adhesives are not directly affected by this regulation but are only indirectly involved through the end product if safety (and usability) of the article depends on the adhesive's suitability.

Adhesives as Toys or Components of Toys for the Purpose of the Toys regulation/Equipment and Product Safety Act - Toys Directive - EN 71
EN 71 is being revised by order of the European Commission. The basis of this review are the safety requirements of the toys directive 88/378/EEC whereby toys, i.e. products which are intended for children up to the age of 14 to play with, must be safe prior to being marketed. Compliance with the requirements of EN 71 is documented by the CE mark. Companies can carry out the test themselves or they can commission a test organization to do so. The EU type approval may be sought if compliance with EN 71 cannot be otherwise determined.

The DIY marketing Committee has developed guidelines showing the steps which need to be taken to demonstrate conformity. You can download the leaflet **"Guideline adhesives and toys"** from the internet at http://www.klebstoffe.com/07_publikation/pdf/TKHHB-MB%201%20-%20Spielzeug.pdf.

Adhesives as Accessories for the Purposes of the Medical Devices act
Adhesives are at the most accessories (for instance in patches) as far as the medical product act is concerned and are therefore not subject to the requirement of a CE mark for medical accessories. Compliance with applicable requirements may nevertheless be requested from manufacturers of medical products. To ensure conformity with the requirements of the European Directive 93/42/EEC, it is important to carry out the requirements laid down in appendix I - general requirements.

The great diversity and variety of medical devices and the levels of complexity and risk involved require that manufacturers must apply the "general requirements" to their specific product. In doing so, the "harmonized standards" of the European Directive are applicable, for instance the set of standards "Biological evaluation of medical devices - Part 1. Evaluation and testing (ISO 10993-1:1997)". The interests of the medical products industry are represented by BVMed (phone +49 (0) 30-2 46 25 50) - www.bvmed.de.

Elements for identifying products in emergency enquiries (DIN EN 15178), identification checklist
The purpose of the product identification standard DIN EN 15178 is to improve product identification in emergency enquiries. The letter "i" near the bar code on packaging refers to the trade

or product name or the number under which the product is registered or officially approved. TRGS 200 the "Classification and identification of substances, preparations and products", dated February 2007, is another guideline for completing an identification checklist which is currently being developed by TKHHB.

Initiatives for Emission Assessment

The committee for assessing the health-related aspects of construction products (AgBB) has issued a product-related assessment method, whereby VOC emissions can be tested using a test chamber. Furthermore, single substances can be measured and assessed, the limit values being so-called "LCI" values (Lowest Concentration of Interest). The current edition of the assessment scheme as well as the related LCI list can be downloaded from http://www.umweltbundesamt.de/bauprodukte/agbb.htm.

Currently DIBt regulations for construction products are limited to Germany and household adhesives are not affected at all.

The European Commission is dealing with the topic of "indoor air" as a high priority and has ordered various projects and studies. By 2010, applicable test standards are to be harmonized within the framework of the construction products Directive (89/106/EEC) by CEN/TC351. At the request of the Executive Board, GEV has offered to apply the EMICODE system to all types of adhesives, including those outside of the area of construction, insofar as the requirements are complied with.

Packaging Regulation

In 2009 the 5th Packaging Act amendment will enter into force. This provides for notification of the quantity of packaging used to the Chambers of Industry and Trade and mainly concerns the manufacturers of products which are "typically" intended for private consumers or "equivalent users", including small craft businesses or freelancers.

Likewise, manufacturers serving industrial customers must provide a certified "declaration of completeness" even if only a small part of their products are "typically" intended for private consumers.

This compulsory registration is intended to combat the declarations made by some manufacturers that they will accept used packaging material themselves, without affiliating to a dual recycling system. Compulsory participation in a dual system and the delivery of the so-called "declaration of completeness" to local Chambers of Industry and Trade is designed to put an end to this situation. Generally it is the "first marketer" who is liable for fees and registration, that is the company which is identified on the label as being the manufacturer responsible (even in the event of private label production).

Unit Pricing/Prepackaging

In Germany, the trade is generally obliged to mark products with basic prices (for instance price per kilogram or litre). This also applies to adhesives, except those below 10 g or ml (§ 9 clause 4 PAngV). However cyano-acrylate adhesives are exempted from the obligation to mark the basic price.

Pursuant to § 7 FPV (Ordinance on Packages), the obligation to quote the filling quantity by weight is still applicable. The European nominal quantities directive 80/232 rules that EU member states must not impose restrictions on filling quantities for products named in its appendices for the purpose of sale or apply limitations in terms of filling quantity, art. 5. The German nominal quantities act (FertigpackungsV) does not provide for restrictions on sales classified by filling quantities.

Technical Committee
Wood and Furniture Adhesives (TKH)

TKH's main area of work is related to consultancy, monitoring a variety of activities from the wood processing arena, establishing scientific journals and working on standards. The areas related to protecting workers, the environment and consumers are gaining increasing importance. In this context, TKH maintains relationships with the German furniture association and major test and approval organizations including the Institut für Fenstertechnik (ift-Rosenheim), the Institut für Holztechnologie (IHD – Dresden) and the Deutsche Institut für Bautechnik (DIBT). Furthermore, close cooperation is maintained with a number of testing houses across Europe.

TKH works on topical issues related to adhesive technology for the wood and furniture industries. They hold round tables of experts and develop leaflets to communicate these topics to users and help them with selecting technical adhesives and use products appropriately. TKH is committed to developing a set of technical briefing notes that reflect the current situation with regard to a number of specific adhesives topics.

As a result of its technical adhesives' know-how, TKH sees itself as a competent contact for testing all kinds of wood adhesive, ranging from PVAc wood adhesive through hot-melt adhesives to reaction adhesives.

Current work by standardization committees dealing with wood and wood material bonds is actively supported. A major task is to monitor activities at a European and international level and to represent national interests.

The committee was successful in lobbying against a proposed standard to control the production of wood scantlings and half-finished sections, and this is now being redrafted.

As new standards appear, TKH publicizes comments on appropriate use of these standards where needed. An example is the revised version of EN 14257, the former WATT 91.

Technical Committee
Self Adhesives Tapes (TKK)

The Technical Committee (TKK) on Self Adhesive Tapes is focused on particular issues related to the European and international standardization of measurement procedures and industry-specific requirements, for instance of the automotive and the electrical industries. Apart from the quality standards ISO TS 16949 already known about, environment-related issues are gaining increasing importance.

In conjunction with organizations in the USA and Japan, methods for measuring adhesive strength, shear resistance and breaking strength, which previously differed from region to region, have now been harmonized and filed by AFERA, the European Association for the Self Adhesive Tape Industry, to the ISO secretary for global application in all countries. Harmonization has meanwhile been completed across Europe (CEN) and at an international level (ISO 29862, ISO 29863 and ISO 29864). Furthermore, the Technical Commission considers the harmonization of unwind adhesion and accelerated ageing to be reasonable. With regard to the use of various rolling ball methods, the Technical Commission aspires to a European overview and assessment system within the framework of AFERA. Another area of work is AFERA's proposed length and width measurements, which are of particular relevance for highly technical products.

Other topics mainly concern the automotive industry:
- Detailed information on product composition must be substantiated in the International Material Data Sheet (IMDS); the Adhesives Association has so far managed to protect our know-how in negotiations between industry associations.
- Design revisions are subject to approval as they are customer requirements; also included are product shifts, changes of raw material suppliers, changes to the process chain in production. The Technical Commission feels that changes are only applicable and need to be communicated when the original product specification is departed from or when function, design or appearance of products are jeopardized by the actual customer application. The members must agree how to handle these changes with their customers. The production acceptance procedure (PPAP to QS 9000 and PPF to VDA) is a comprehensive documentation containing a broad range of risk analyses, control and process plans and certificates of proficiency for test and process equipment which self adhesive tape manufacturers must submit for customer approval. PPAP/PPF documents contain a huge amount of know-how so that appropriate measures to protect confidentiality must be ensured while avoiding negligence of customer interests.

The work of the Technical Committee supports AFERA. The Technical Committee contributes to the "AFERA News" brochure with reports on meetings and events of the adhesive tape industry and new products.

Technical Committee
Paper and Packaging Adhesives (TKPV)

Adhesives for Food Contact Materials and Articles:

The Technical Committee on Paper and Packaging Adhesives focused its activities on continuing work related to assessing adhesives for foodcontact materials and articles. It is still the case that adhesives for producing materials and articles intended to come into contact with food are not specifically governed by European regulations. Adhesives as part of foodcontact materials and articles are however subject to being assessed under a food law (framework regulation (EC) 1935/2004). When adhesives' formulations based on polymers and additives are used to produce plastic for materials and articles intended to come into contact with food, the related appendices to this regulation also apply to them. In all other cases, national regulations such as the recommendations of the Federal Institute of Risk Assessment (BfR) remain valid.

The European Commission responsible for legislation on materials and articles intended to come into contact with food has discussed how far adhesives can be covered by European directives. In appendix 1 to the new framework regulation (EC) 1935/2004, adhesives are named in a list of materials and articles for which specific regulations shall be developed. Exactly what a specific regulation for adhesives might look like is still subject for discussion. Besides the possibility of a positive list as is used in other specific regulations, a guide to "good manufacturing practice" or a "practical guide" for applying adhesives to materials and articles intended to come into contact with food might be options.

The toxicological analysis which is required for each individual substance to be included in the positive list will be quite laborious and expensive. In particular, there is concern within the field of additives, auxiliary materials etc. that testing certain (potentially slow selling) products will mean that they will be excluded, which would result in them no longer being available to the adhesives industry as approved raw materials. This issue has also been considered by the European Commission. For this reason, a practical guide for the use of adhesives is the favoured option. The introduction of a practical guide would be reasonable as adhesives play a very limited role in most packaging applications and generally there are barrier layers between the adhesive and the food. The introduction of a practical guide would render a positive list unnecessary.

Under the leadership of TKPV, IVK, together with IVV (Fraunhofer Institut für Verfahrenstechnik & Verpackung), has analyzed the feasibility of such a practical guide. For the required scientific tests, a European project has been prepared. From the point of view of the industry, the European project "MIGRESIVES II", which was launched in February 2007, aims at developing a reliable and economical model for calculations so that adhesive formulations and applications can be adapted to ensure safe use. This makes expensive measurements often unnecessary. The conditions of grant specify that the organization, coordination, financing and execution of the project lies exclusively in the hands of small and medium-sized enterprises and associations. For this reason, two medium-sized members of the consortium, who are also represented in TKPV, represent the interests of TKPV as a whole. IVK as one of the project partners ensures that raw material producers and multinational corporations are involved within the boundaries of what is legally possible.

In order to improve communication throughout the supply chain, TPKV has developed a briefing note which is a type of "Declaration of Legal Status" for adhesive raw materials and adhesives. This paper describes the legal status of adhesives and adhesive raw materials from the perspective of the food law, as no formal declaration of conformity can be prepared due to the lack of a specific adhesives directive. In order to include foreign raw materials manufacturers in particular, TKPV has translated the TKPV 1 briefing note" Food law status of adhesives and adhesives raw materials" into English.

With the establishment of the Commission Regulation (EC) No. 2023/2006 on good manufacturing practice for materials and articles intended to come into contact with food" a regulation has meanwhile come into effect which substantiates the idea of good manufacturing practice as is requested in article 3 of Regulation No. 1935/2004 on materials and articles intended to come into contact with food. For the field of adhesives and adhesive raw materials, TKPV has developed the "Good Manufacturing Practice" guidelines for adhesives used to produce foodcontact materials and articles intended to come into contact with food (TKPV briefing note no. 4).

Another EU initiative in the context of evaluating articles intended to come into contact with food is the "FACET" project. FACET (Flavours, Additives and Contact materials Exposure Task) deals with the development of scientifically based tools to enable the effects of exposure of the human body to chemicals, for instance those contained in materials which come into contact with food, as well as in food to be evaluated. (Exposure is determined among other things by examining which packaging material is used for which food, how packaging is designed and what packaging quantities are used.) As adhesives are part of the packaging, IVK supports the European FEICA association in their commitment to the FACET project. TKPV will log relevant data for the German market.

Adhesives in Paper Recycling:
Another major area of TKPV's work involves the impact of adhesives on paper recycling. The agreement negotiated with the graphics industry, the European Commission and the Federal Ministry for the Environment, Nature Conservation and Nuclear Safety implies that higher quantities of paper must be recycled.

The "European Declaration on Paper Recycling" meets this requirement. This declaration has been signed by a range of organizations including the European Adhesives Organisation (FEICA). TKPV is involved in an ongoing, intensive dialogue with the paper industry and scientific institutions on this topic. A number of scientific events have been held related to this topic, and TKPV has presented their position on the issue of "adhesives contamination".

IVK has issued a status report on the activities of TKPV and the various association members, which is included in the publication on the "European Declaration on Paper Recycling".

As a result of both foodstuff and non-foodstuff packaging being mixed in recycling, together with the use of DIBP as a softener for homopolymer dispersions, for instance in adhesives for non-foodstuff packaging, high levels of DIBP are today found in recycled paper. In order to be able to use recycled fibres for food packaging in the future, the use of DIBP should be largely avoided throughout the paper production chain. TKPV therefore recommends that IVK members cease offering adhesives containing DIBP to the paper and board producing or processing industry as of 1 January 2009. This recommendation has been successfully put into action.

This issue is of particular importance as for materials and articles intended to come into contact with foodstuffs, DIBP was removed from the list of additives and monomers for plastic approved for contact with food in 2007 (annex III of directive 2002/72/EC). For all other areas, the classification of being toxic to reproduction, category 2, and identification of preparations containing 25 % DIBP or more with T, R61 will be applicable in future (presumably from 1 June 2009 onwards).

REACh:
TKPV is involved with activities dealing with the new European chemicals regulation "REACh" (Registration, Evaluation, Authorisation and Restriction of Chemical substances).

In April 2008, during the course of the TKPV symposium "Adhesives in the Paper and Packaging Industry", a competition-neutral forum for exchanging information, REACh was discussed. A number of speakers addressed the complex effects for customers of companies affiliated with TKPV under the banner „REACh – Consequences for the Paper Processing and the Packaging Industry".

Currently TKPV is working on exposure scenarios in the field of paper and packaging applications for adhesives.

Standardization:
TKPV's has expert members in a range of committees and workgroups, such as the workgroup "Adhesives for Paper, Board, Packaging and Disposable Sanitary Products" of the European standardization committee CEN/TC 193 and in the national toys committee DIN/NMP 458. They successfully continue to represent the technical aspects of German paper and packaging adhesives.

FEICA:
Due to their importance at a European level, the topics „adhesives for materials and articles intended to come into contact with food", "MIGRESIVES II", "FACET", "GMP" and "Recycling" are, on the initiative of TKPV, being intensely discussed with the European „Paper and Packaging" workgroup of FEICA. The aim is to come to a European position on the adhesives industry and to find solutions to these important issues. The sanitary guidelines established by TKPV as well as the leaflet "Food law status of adhesives and adhesive raw materials" have therefore been translated into English and made available to the European adhesives industry.

Other topics dealt with:
STLB label adhesives for spirits bottles , voluntary labelling of natural rubber latex

Technical Committee
Footwear and Leather Adhesives (TKS)

TKS coordinates the technical public relations of German manufacturers of adhesives for footwear and leather materials. In addition, the commission supports national and international standardization activities and acts as the point of contact for market segment-specific technical evaluations and information within the activities of IVK and German Chemical Industry Federation (VCI) as well as for regulatory affairs.

Work is focused on cooperating in the development of national and European standards to define the basic properties of footwear and leather adhesives.

Activities include standards for
- Minimum requirements for footwear bonds (requirements and materials)
- Tests of the resistance of footwear bonds (peel resistance tests)
- Processing (determination of optimum activation conditions, determination of sole positioning tack)
- Resistance (colour change by migration, thermal resistance of lasting adhesives)

The standards defining the minimum requirements and testing of footwear bonds are currently being revised, and it is planned to develop a uniform classification concerning the minimum resistance requirements, an extension to cover new sole materials and a more detailed description of all ageing tests.

In order to be able to carry out a large number of tests, the provision and availability of standardized reference test materials and test adhesives is inevitable. Therefore the selection and specification of test materials and adhesives are verified and updated on a continuous basis.

Training
Training in adhesives for employees in the footwear industry is another task of TKS.

TKS seeks to share the adhesive-specific know-how necessary so as to be able to quickly identify practical problems, work on solutions and implement substantiated measures, thus minimizing and even eliminating in the future potential error sources.

In this context, TKS has developed a training course which was first offered in conjunction with IFAM/Bremen and PFI/Pirmasens in the form of a practical seminar named "Applied adhesive technology in the footwear industry". More than 20 trainees attended the seminar. Following the positive feedback received, this seminar is now repeated every year.

This seminar focuses on practical examples showing optimal adhesive application and typical errors. The attendees are then able to apply what they have learned in exercise sessions. Theoretic background knowledge thus becomes understandable and transparent for practitioners working in production.

The third TKS seminar is scheduled to take place in spring 2009.

Technical Committee
Structural Bonding and Sealing (TKSKD)

The newest technical committee within IVK supports the related working committee by acting as a forum for focused work on technical issues, thereby helping the manufacturers of structural adhesives and sealants as well as corresponding research institutes.

Structural adhesives and sealants are used for instance in transport (car, train and aircraft production and shipbuilding), the electric and electronics industry, medical technology and optical industry as well as in mechanical engineering, plant and equipment engineering.

Work so far has focused on, for instance.
* Providing information to IVK members about the IMDS system initiated by the automotive industry to characterize the materials used in cars and promoting active participation in the development and shaping of this system. Having successfully established adhesives and sealants as standard materials promoted by the IMDS steering committee, the next step will be to define self adhesive tape as a standard material. Furthermore, a "Europeanization" of IMDS can be observed.
* Quality control of adhesives training. IVK is contributing towards the funding of a training project initiated by IFAM (internationalization of adhesives engineer training).
* Standardization. The members of the structural adhesives standardization committee financed by IVK report on a regular basis on standardization activities at a national and a European level. Besides developing recommendations for IVK members concerning the CE mark for structural adhesives (prEN 15275), TKSKD has among other things supported the initiative of the American Adhesive and Sealant Council (ASC) to establish a standard for characterizing bonds in CAD systems. The standard will first be published in the USA as an ASTM standard and is then to be carried over to an ISO standard. Thanks to early involvement in this project, German and European interests have been taken into account.
TKSKD is also dealing with research funding in the field of adhesives and the identification of potential pre-competitive research projects.

REACh will be a major focus in the near future. In addition to information provided to member companies on the current activities of various organizations (FEICA, TB, VCI, etc,), the TC members will join in with the development of structural adhesive-specific standardized phrases for exposure scenarios to be established by adhesives manufacturers.

The workgroup also sees itself as the bridge between industry and extra-industrial scientific activities.

Public Relations Committee (BeifÖ)

The key task of the Public Relations Committee is to present the German Adhesives Association and the key technology of adhesives bonding to the general public.

The ongoing public relations work of the German Adhesives Association is bearing fruit: Press coverage of IVK during 2007 was very well received across a range of media, including trade journals, daily newspapers and magazines, as can be seen by the compiled press cuttings: The print run and thereby the number of readers reached is steadly increasing. The type of documentation was also improved. In addition to the print run, an analysis was made of equivalent advertising prices.

These days ever more editorial departments, whilst still using the topics of regular press information, try to get to the bottom of the topic and provide detailed coverage, not only in the print media but also on the radio, TV and on the internet.

The internet press platform www.klebstoff-presse.com which was designed to meet the special requirements of journalists had a relatively high hit rate over the past year. The same applied to the internet portal www.klebstoffe.com.

The press work of the association is currently complemented by a new newsletter for members, as well as the newsletter "Kleben fürs Leben".

This first appeared for the 60th anniversary of the German Adhesives Association in 2006, and literally sold out within a few weeks. The press also received copies of this publication and of publicized reports which originally appeared in "Kleben fürs Leben": One year later, the second edition was published with an increased print run. It was also well received. 2008, an English version was presented at the World Adhesives Conference held in Miami, Florida, and a third German newsletter was also issued, thus increasing in a very important and invaluable way the role of the Public Relations Committee.

Carlo Jahr, Wacker Chemie, was elected new spokesman of the PR committee. He took over the role from Wolfgang Stauf who had become unable to continue this function for professional reasons.

IVK Management

The employees of the German Adhesives Association office take care of coordinating, handling and following up on the tasks which result from various committees. The office keeps members up to date on new topics and serves as an information exchange for the association members.

The management of IVK also perceives itself as a representative and competent partner to the adhesives industry. Within this role, the management represents the technical and economic interests of the industry with national, European and international authorities, customer and consumer associations, system partners, institutions and the public. Through active involvement on the advisory boards of major trade shows and trade journals, advisory committees of the German Parliment and working committees of various Federal and European ministries, the management keeps track of adhesives-related topics and projects from an expert and a content perspective.

This also applies to research – in its role as a board member of the DECHEMA's (society for Chemical Engineering and Biotechnology) "Adhesives Technology" division, the management team helps with coordinating publicly-funded scientific research projects in the field of adhesives and adhesives technology – and contributes to the industry-wide dialogue which IVK initiated a few years ago.

The management team produces a range of publications, holds discussions with interested circles and gives lectures and teaching assignments. The end result is effective communication about the broad range of services and the high potential for innovation of the adhesives industry as well as the exemplary commitment of its members to protecting workers, consumers and the environment.

Committees of the
German Adhesives Association (IVK)

Executive Board

Chair of the Board: Dr. Ralf Schelbach	JOWAT AG Beijing P. R. China
Deputy Chair of the Board: Dr. Joachim Schulz	EUKALIN Spezial-Klebstoff Fabrik GmbH D-52249 Eschweiler
Other Board Members:	
Klaus Becker-Weimann	Kleiberit Klebstoffe Klebchemie M. G. Becker GmbH & Co. KG D-76356 Weingarten
Dr. Jochen Billecke	ARDEX GmbH D-58453 Witten
Dirk Brandenburger	Sika Automotive GmbH D-22525 Hamburg
Hans-Detlef Deike	Bostik GmbH D-33829 Borgholzhausen
Dieter A. Gundlach	ARDEX GmbH D-58453 Witten
Ansgar van Halteren	Industrieverband Klebstoffe e.V. D-40219 Düsseldorf
Dr. Achim Hübener	Henkel AG & Co. KGaA D-69112 Heidelberg
Klaus Kullmann	Jowat AG D-32758 Detmold
Dr. Thomas Pfeiffer	Türmerleim GmbH D-67061 Ludwigshafen
Peter Rambusch	certoplast Vorwerk & Sohn GmbH D-42285 Wuppertal

Reimond Rohde	Exxon Mobil Chemical Central Europe GmbH D-50735 Köln
Dr. H. Werner Utz	UZIN UTZ Aktiengesellschaft D-89079 Ulm
Dr. Hubertus von Voithenberg	The ChemQuest Group, Inc. D-22609 Hamburg
Andreas Walkembach	tesa AG D-22525 Hamburg

Technical Board

Chair of Technical Board: Dr. Achim Hübener	Henkel AG & Co. KGaA D-69112 Heidelberg

Other Board Members:

Reiner Asdonk	National Adhesive GmbH D-47533 Kleve
Dr. Peter Broschk	tesa AG D-20253 Hamburg
Dr. Bernd Burchardt	Sika Services AG CH-8967 Widen, Schweiz
Franz Diehlmann	Weiss Chemie + Technik GmbH & Co. KG D-35708 Haiger
Stefan Endisch	BASF SE D-67056 Ludwigshafen
Dr. Frank Gahlmann	Stauf Klebstoffwerk GmbH D-57234 Wilnsdorf
Dr. Knut Göke	Kömmerling Chemische Fabrik GmbH D-66954 Pirmasens

Prof. Dr. Andreas Groß	IFAM Fraunhofer-Institut für Fertigungstechnik und Angewandte Materialforschung D-28359 Bremenn
Dirk Lamm	tesa AG D-22525 Hamburg
Dr. Hartwig Lohse	Klebtechnik Dr. Hartwig Lohse e. K. D-25524 Itzehoe
Dr. Heinz-Werner Lucas	Bayer MaterialScience AG D-51368 Leverkusen
Dr. Hermann Onusseit	Henkel AG & Co. KGaA D-40191 Düsseldorf
Matthias Pfeiffer	Türmerleim GmbH D-67061 Ludwigshafen
Timm Schulze	Jowat AG D-32758 Detmold
Dr. Christian Terfloth	Jowat AG D-32758 Detmold

Technical Committee Building and Flooring Adhesives

Chair: Dr. Frank Gahlmann	Stauf Klebstoffwerk GmbH D-57234 Wilnsdorf

Other Members:

Dr. Norbert Arnold	UZIN UTZ Aktiengesellschaft D-89079 Ulm
Dr. Jürgen Bezler	WACKER Polymers GmbH & Co. KG D-84489 Burghausen
Dr. Thomas Brokamp	Bona GmbH Deutschland D-65549 Limburg

Dr. Alexander Centner	BASF SE
D-67056 Ludwigshafen
Jürgen Gehring	Bostik GmbH
D-33829 Borgholzhausen
Dr. Matthias Hirsch	Kiesel Bauchemie GmbH u. Co. KG
D-73730 Esslingen
Michael Illing	Forbo Erfurt GmbH
D-99091 Erfurt
Dr. Hubert Motzet	Schönox GmbH
D-48720 Rosendahl
Berhard Mucherl	Murexin AG
A-2700 Wiener Neustadt
Dr. Rüdiger Oberste-Padtberg	ARDEX GmbH
D-58453 Witten
Helmut Schäfer	Sopro Bauchemie GmbH
D-65203 Wiesbaden
Dr. Martin Schäfer	Wakol
Walter Kolodziej GmbH & Co. KG
D-66954 Pirmasens
Hartmut Urbath	Henkel AG & Co. KGaA
D-40589 Düsseldorf

Technical Committee Wood and Furniture Adhesives

| Chair:
Timm Schulze | Jowat AG
D-32758 Detmold |
|---|---|

Other Members:

| Dr. Horst Beck | Henkel AG & Co. KGaA
D-40191 Düsseldorf |
|---|---|

Holger Brandt	Follmann & Co. GmbH & Co. KG D-32423 Minden
Dr. Wilhelm Eib	Zika Leime + Klebstoffe L. Zimmermann GmbH + Co. KG D-72793 Pfullingen
Daniela Hardt	Celanese Emulsions GmbH D-65926 Frankfurt
Oliver Hartz	BASF SE D-67056 Ludwigshafen
Jürgen Lotz	Henkel KGaA D-73442 Bopfingen
Dr. Oliver Pankow	Wacker Chemie AG D-84489 Burghausen
Dr. Werner Praß	Türmerleim GmbH D-67061 Ludwigshafen
Dipl.-Ing. Martin Sauerland	Kleiberit Klebstoffe Klebchemie M. G. Becker GmbH & Co. KG D-76356 Weingarten
Walter Stadlbauer	H.B. Fuller Deutschland GmbH D-31582 Nienburg
Dr. Claudia Yaacoub	Henkel AG & Co. KGaA D-73442 Bopfingen

Technical Committee DIY and Consumer Adhesives

Chair:
Dr. Dirk Lamm

tesa AG
D-20253 Hamburg

Other Members:

Frank Avemaria

3M Deutschland GmbH
D-41453 Neuss

Dr. Christine Egenhofer-Breuer

3M Deutschland GmbH
D-41453 Neuss

Dr. Wolfgang Maier

Henkel AG & Co. KGaA
40191 Düsseldorf

Technical Committee Self Adhesive Tapes

Chair:
Dr. Peter Broschk

tesa AG
D-20253 Hamburg

Other Members:

Dr. Andreas Hohmann

certoplast Vorwerk & Sohn GmbH
D-42285 Wuppertal

Lutz Jacob

Jacob Industrie Consulting
D-87527 Altstaedten

Harald Kurth

Lohmann GmbH & Co. KG
D-56567 Neuwied

Christoph Lodde

COROPLAST Fritz Müller GmbH & Co. KG
D-42279 Wuppertal

Thomas Müller

Kömmerling Chemische Fabrik GmbH
D-66954 Pirmasens

Jürgen Peters	3M Deutschland GmbH D-41453 Neuss
Dr. Ralf Rönisch	Henkel AG & Co. KGaA D-40191 Düsseldorf
Helge Schönfeld	Sika Automotive GmbH D-22525 Hamburg
Johannes Türk	BASF SE D-67056 Ludwigshafen

Technical Committee Paper and Packaging Adhesives

Chair: Dr. Hermann Onusseit	Henkel AG & Co. KGaA D-40191 Düsseldorf

Other Members:

Ralf Ardner	Follmann & Co. GmbH & Co. KG D-32423 Minden
Carsten Branding	Friedrich Branding GmbH & Co. KG EFBECOL-Klebstoff-Fabrik D-31275 Lehrte
Dr. Manfred Dietz	Klebstoffwerke COLLODIN GmbH & Co. KG D-60386 Frankfurt
Franz-Heinrich Eiden	Bostik GmbH D-33829 Borgholzhausen
Dr. Rainer Fasolt	Planatol Adhesive GmbH D-83101 Rohrdorf-Thansau

Hans Joachim Fricke	BASF SE D-67056 Ludwigshafen
Dr. Ulrich Holtin	H.B. Fuller Deutschland GmbH D-21335 Lüneburg
Dr. Gerhard Kögler	Wacker Chemie AG D-84489 Burghausen
Bernhard Kohlhoff	Jowat AG D-32758 Detmold
Dr. Bernhard Momper	Celanese Emulsions GmbH D-65926 Frankfurt
Matthias Pfeiffer	Türmerleim GmbH D-67061 Ludwigshafen
Arno Prumbach	EUKALIN Spezial-Klebstoff Fabrik GmbH D-52249 Eschweiler
Hans-Günter Schetschok	Rohm and Haas Deutschland Produktion GmbH & Co. KG D-28309 Bremen

Technical Committee Footwear and Leather Adhesives

Chair: Dr. Knut Göke	Kömmerling Chemische Fabrik GmbH D-66954 Pirmasens

Other Members:

Wolfgang Arndt	Bayer MaterialScience AG D-51368 Leverkusen
Martin Breiner	Kömmerling Chemische Fabrik GmbH D-66954 Pirmasens
Andreas Ecker	H.B. FULLER Austria GmbH A-4600 Wels, Österreich

Oliver Hartz	BASF SE D-67056 Ludwigshafen
Dr. Roland Heider	Adtracon GmbH D-40721 Hilden
Dr. Martin Schäfer	Wakol Walter Kolodziej GmbH & Co. KG D-66954 Pirmasens

Technical Committee Structural Bonding and Sealing

| Chair:
Dr. Hartwig Lohse | Klebtechnik Dr. Hartwig Lohse e. K.
D-25524 Itzehoe |

Other Members:

Martin Breiner	Kömmerling Chemische Fabrik GmbH D-66954 Pirmasens
Franz Diehlmann	Weiss Chemie + Technik GmbH & Co. KG D-35708 Haiger
Gerd-Rolf Dietz	Henkel AG & Co. KGaA D-69112 Heidelberg
Karl Engeldinger	tesa AG D-20253 Hamburg
Prof. Dr. Andreas Groß	IFAM Fraunhofer-Institut für Fertigungstechnik und Angewandte Materialforschung D-28359 Bremen
Dr. Karl Michael Müller	Bostik GmbH D-33829 Borgholzhausen
Dr. Manfred Schumann	Henkel AG & Co. KGaA D-69112 Heidelberg
Artur Zanotti	Sika Deutschland GmbH D-72574 Bad Urach

Public Relations Committee

Spokesperson: Carlo Jahr	Wacker Polymers GmbH & Co. KG D-50769 Köln
Other members: Rolf J. Blaas	ROHM AND HAAS EUROPE TRADING APS Deutsche Zweigniederlassung D-60486 Frankfurt
Thomas Dülberg	Dülberg & Brendel GmbH D-40211 Düsseldorf
Ansgar van Halteren	Industrieverband Klebstoffe e. V. D-40219 Düsseldorf
Oliver Jüntgen	Henkel AG & Co. KGaA D-40191 Düsseldorf
Peter Kreft	Henkel AG & Co. KGaA D-40191 Düsseldorf
Michael Kriesten	Wetzel GmbH & Co. KG D-32052 Herford
Jan Szöts	Gludan (Deutschland) GmbH D-21514 Büchen

Marketing Committee Building and Flooring Adhesives

Chair: ARDEX GmbH
Dieter A. Gundlach   D-58453 Witten

Marketing Committee Wood and Furniture Adhesives

Chair: Kleiberit Klebstoffe
Klaus Becker-Weimann   Klebchemie M. G. Becker GmbH & Co. KG
                                D-76356 Weingarten

Marketing Committee on Industrial Adhesives

Chair: Jowat AG
Dr. Ralf Schelbach   Beijing
                                P. R. China

Marketing Committee Paper and Packaging Adhesives

Chair: Türmerleim GmbH
Dr. Thomas Pfeiffer   D-67061 Ludwigshafen

Marketing Committeee Raw Materials

Chair: Exxon Mobil Chemical Central Europe GmbH
Reimond Rohde   D-50735 Köln

Marketing Committee Self Adhesives Tapes

Chair: certoplast Vorwerk & Sohn GmbH
Peter Rambusch   D-42285 Wuppertal

Marketing Committee Structural Bonding and Sealing

Chair: Sika Automotive GmbH
Dirk Brandenburger   D-22525 Hamburg

Marketing Committee DIY and Consumer Adhesives

Chair:
Dirk Lamm

tesa AG
D-22525 Hamburg

Marketing Committee Footwear and Leather Adhesives

Chair:
Martin Breiner

Kömmerling Chemische Fabrik GmbH
D-66954 Pirmasens

Marketing Committee Foam Adhesives

Chair:
Hermann Weidmann

Henkel AG & Co. KGaA
D-73442 Bopfingen

IVK-Management

Danuta Dworaczek
Ansgar van Halteren
Dr. Axel Heßland
Ingrid Peschges
Michaela Szkudlarek
Martina Weinberg
Klaus Winkels

Honorary Members

Dr. Johannes Dahs
Dr. Hannes Frank
Otto A.Stein
Dr. Rainer Vogel
Dipl.-Chem. Heinz Zoller

Königswinter
Detmold
Pirmasens
Langenfeld
Pirmasens

Holders of the Medal of the German Adhesives Industry

Dr. Hannes Frank
Prof. Dr. Otto-D. Hennemann

Detmold
Osterholz-Scharmbeck

Medal of the German Adhesives Industry

In 2007 the Executive Board and the General Assembly of the German Adhesives association decided to award the Medal of the German adhesives industry to two individuals for their outstanding service to the adhesives industry and adhesives technology.

The medal was awarded to

Dr Hannes Frank
in recognition of decades of commitment to the German adhesives industry. As a member of the technical committee, Dr Frank has greatly promoted and helped shape adhesives technology and the image of the adhesives industry, for example he has been very committed to medium-sized companies and their potential to innovate, which is indispensable for technical and economic development. In addition to that, Dr Frank is a successful pioneer of polyurethane adhesives technology. Furthermore, he has promoted an industry-wide strategy on communication and training, thereby contributing to making adhesive bonding a key technology of the 21st century.

Prof. Dr Otto-Diedrich Hennemann
in recognition of his scientific work, mainly promoting and shaping the "bonding system". This includes his research into the long-term resistance of bonds and the implementation of appropriate simulation processes in the automotive and aerospace industries. His research methodology was to always focus on concrete applications and the development of added value for system partners.

FEICA (Fédération Européenne des Industries de Colles et Adhésifs - Association of European Adhesives and Sealants Manufacturers) has been representing the interests of the European adhesives industry since 1972. FEICA is the umbrella organization for 14 national adhesives associations across Europe, currently representing the interests of eight multinational adhesives manufacturers.

From 1972 until late 2006 FEICA shared an office and was jointly managed with the German Adhesives Association at its headquarters in Dusseldorf. Then, in 2007, FEICA moved to Brussels.

FEICA'S Objectives

FEICA works with its members to represent the combined interests of the adhesives industry in Europe. It is particularly committed to representing the interests of its members with the institutions of the European Union.

FEICA plays a leading role in the adhesives industry with regard to the harmonization of national and European legislation and in the development of European standards.

External Relations

In order to be constantly up to date with projects and decrees by the European Union (European Parliament, European Council, European Commission, General Directorates), FEICA has close relationships with CEFIC (European Chemical Industry Council) and thereby to other European associations (e. g., Plastic Europe).

Within the framework of European Standardization, cooperation with CEN (Comité Européen de Normalisation, European Committee for Standardization) plays a special role, in particular work with the "Technical Committee" TC 193 "adhesives", whose meetings FEICA attends on a regular basis.

Service for Members

As FEICA's largest member, the German Adhesives Association (IVK) represents the interests of German adhesives manufacturers on the Executive Board, the European Technical Board as well as various other technical committees of the European association. This ensures that members of the German Adhesives Association receive complimentary services, all the information they need, as well as contact with decision makers at a European level.

Contact Address

FEICA - Av Van Nieuwenhuyse; 6 - B - 1160 Brussels - www.feica.com

Association for the Control of Emissions
in Products for Flooring Installation,
Adhesives and Building Materials

EMICODE

Protection against indoor air pollution

The topic of protection against indoor air pollution and the demand made by key consumers to provide low emission products led, in February 1997, to the foundation of the Association for the Control of Emissions in Products for Flooring Installation Adhesives and Building Materials (GEV) and the introduction of the EMICODE classification system. The initiative was supported by the leading manufacturers of flooring products of the German Adhesives Association.

Early in the 1990s the German adhesives industry in conjunction with the employers' liability insurance association developed the so-called GISCODE to assist traders with selecting the most appropriate flooring material. This initiative was based on the hazardous material law which requests that craftsmen use the least dangerous product. GISCODE provides them with a quick overview for any given product so that they can assess whether it is suitable from a labour protection perspective.

This contributed to a considerable reduction in solvent-based products over the past few years (see illustration).

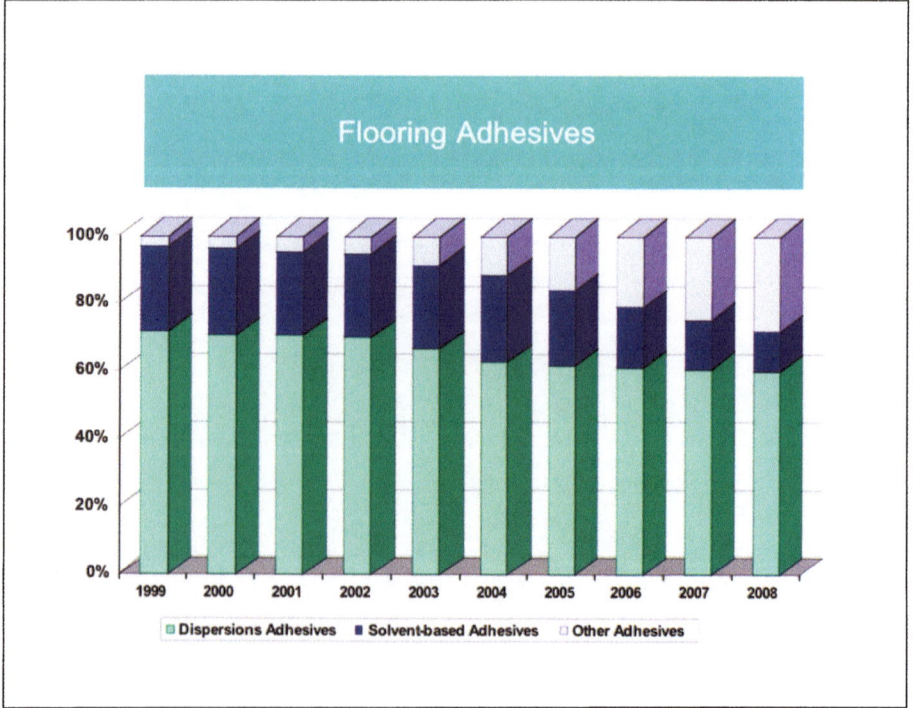

Thanks to this law, parquet and flooring technicians should only be exposed to high concentrations of volatile solvents in exceptional cases which are defined by an industry-specific technical rule applicable to hazardous materials (TRGS 610). The use of substitutes with a much lower hazard potential has therefore become general practice.

However organic compounds which can be described as volatile residues are not sufficiently covered by these provisions. These can be impurities in raw materials which cannot be avoided for technical reasons or low-volatile compounds. These are emitted to the room air in low concentrations, but over a prolonged period of time.

Leading manufacturers of flooring materials have therefore developed a new generation of solvent-free, very low emission flooring materials which are particularly recommendable from the point of view of indoor air hygiene. Meanwhile GEV has more than 50 members in 10 countries and continues to grow, above all in Europe.

In order to provide a reliable guideline for traders and consumers given the plethora of measuring methods, EMICODE was developed. This is a product classification system which is neutral in terms of effect on competition. Using EMICODE, flooring materials can be assessed and their emission behaviour compared. At the same time, it provides a strong incentive to constantly improve products.

The EMICODE classification system is based on a precisely defined test chamber analysis and demanding classification criteria. Adhesives, levelling compounds, primers, sealants, underlays, tapes and parquet coatings which are identified as "very low emission" and bear the GEV mark EMICODE EC 1, offer maximum protection against indoor air pollution and unpleasant smells.

This voluntary initiative is an ongoing effort to protect the health of traders and consumers. EMICODE provides specifiers as well as architects, planners, craftsmen, building owners and Technical documents as well as the articles of association can all be found on the website www.emicode.com.

By including joint sealants amongst the products covered by EMICODE, GEV has complied with a market need for more products, which are not classic flooring products, to be classified. The association responds to the need for ecological aspects to be considered in product classification by including surface treatment products for parquet. For products which require protective measures when used, even though these products are low emission, GEV has introduced the additional letter "R" in the logo (= regulated).

Furthermore, a product database was established with the support of the Federal Environment Agency (www.emicode.com), where member companies upload details of their products in order to provide traders with the required information. This initiative helped to consolidate the large number of complex databases into one central system which is managed by the companies.

CEO   Rüdiger Reinfelder
Head of Technical Advisory Board   Jürgen Gehring
Managing Director GEV   Klaus Winkels, Lawyer

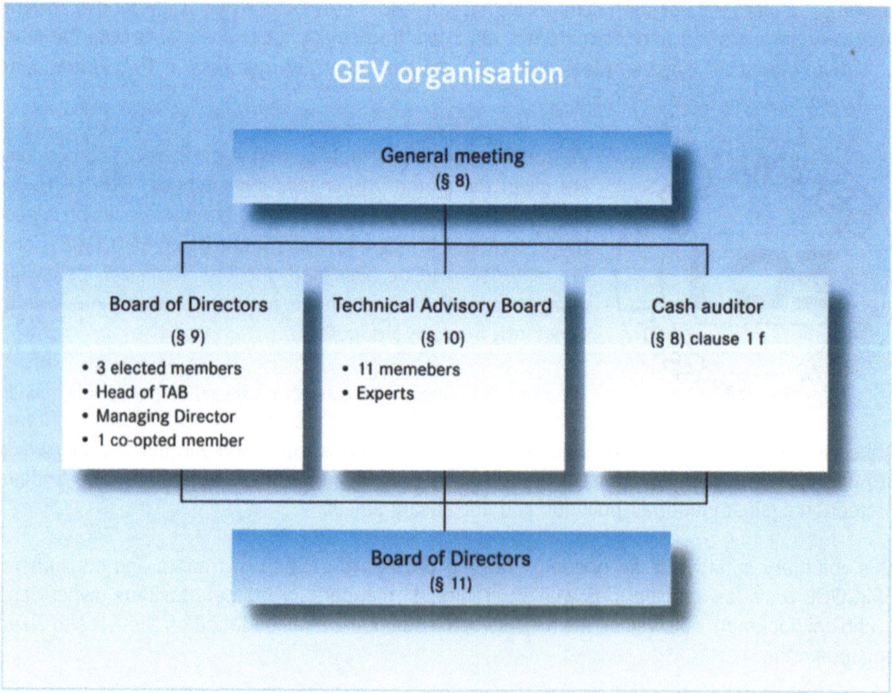

GEV organisation

General meeting
(§ 8)

Board of Directors
(§ 9)
- 3 elected members
- Head of TAB
- Managing Director
- 1 co-opted member

Technical Advisory Board
(§ 10)
- 11 memebers
- Experts

Cash auditor
(§ 8) clause 1 f

Board of Directors
(§ 11)

Gemeinschaft Emissionskontrollierte Verlegewerkstoffe, Klebstoffe und Bauprodukte (GEV) (Association for the Control of Emissions in Products for Flooring Installation, Adhesives and Building Materials)
RWI-House
Völklinger Straße 4
D-40219 Düsseldorf
Phone +49 (0) 2 11-6 79 31-20
Fax +49 (0) 2 11-6 79 31-33
Email: info@emicode.com
www.emicode.com

Relevant Laws and Regulations
for Adhesives

Hazardous Substances

- Regulation (EC) No 1907/2006 of the European Parliament and of the Council of 18 December 2006 concerning the Registration, Evaluation, Authorisation and Restriction of Chemicals (REACH), establishing a European Chemicals Agency, amending Directive 1999/45/EC and repealing Council Regulation (EEC) no. 793/93 and Commission Regulation (EC) no. 1488/94 as well as Council Directive 76/769/EEC and Commission Directives 91/155/EEC, 93/67/EEC, 93/105/EEC and 2000/21/EC
- Directive 2006/121/EC (...) amending Council Directive 67/548/EEC on the approximation of laws, regulations and administrative provisions relating to the classification, packaging and labelling of dangerous substances in order to adapt it to REACH
- REACH fees regulation – Commission regulation (EC) No 340/2008 of 16 April 2008 on the fees and charges payable to the European Chemicals Agency pursuant to Regulation (EC) no. 1907/2006 of the European Parliament and of the Council on the Registration, Evaluation, Authorisation and Restriction of Chemicals (REACH)
- Council directive 67/548/EEC (...) relating to the classification, packaging and labelling of dangerous substances; as amended
- Council directive 1999/45/EC (...) relating to the classification, packaging and labelling of dangerous preparations
- Council directive 76/769/EEC (...) relating to restrictions on the marketing and use of certain dangerous substances and preparations, as amended - repealed as at 1 June 2009
- Council Directive 98/24/EC (...) on the protection of the health and safety of workers from the risks related to chemical agents at work
- Regulation (EC) no.304/2003 of the European Parliament and of the Council concerning the export and import of dangerous chemicals

Waste Legislation

- Regulation (EC) no. 1013/2006 of the European Parliament and of the Council of 14 June 2006 on shipments of waste
- Commission Regulation (EC) no. 1379/2007 of 26 November 2007 amending Annexes IA, IB, VII and VIII of Regulation (EC) No. 1013/2006 of the European Parliament and of the Council on shipments of waste, for the purposes of taking account of technical progress and changes agreed under the Basel Convention
- Commission Decision 2000/532/EC replacing Decision 94/3/EC establishing a list of wastes pursuant to Article 1(a) of Council Directive 75/442/EEC on waste and Council Decision 94/904/EC establishing a list of hazardous waste pursuant to Article 1(4) of Council Directive 91/689/EEC on hazardous waste

Immission Protection Legislation

- Council Directive 1999/13/EC (...) on the limitation of emissions of volatile organic compounds due to the use of organic solvents in certain activities and installations
- Directive 2004/42/CE of the European Parliament and of the Council of 21 April 2004 on the limitation of emissions of volatile organic compounds due to the use of organic solvents in certain paints and varnishes and vehicle refinishing products and amending Directive 1999/13/EC
- Regulation (EC) no. 761/2001 allowing voluntary participation by organisations in a Community eco-management and audit scheme (EMAS)
- Directive 2008/1/EC of the European Parliament and of the Council of 15 January 2008 concerning integrated pollution prevention and control
- Regulation (EC) no. 166/2006 of the European Parliament and of the Council of 18 January 2006 concerning the establishment of a European Pollutant Release and Transfer Register and amending Council Directives 91/689/EEC and 96/61

Water Legislation

- Directive 2008/1/EC of the European Parliament and of the Council of 15 January 2008 concerning integrated pollution prevention and control
- Directive 2000/60/EC of the European Parliament and of the Council establishing a framework for Community action in the field of water policy

Hazardous Materials Transportation Law

- Council Directive 96/49/EC of 23 July 1996 on the harmonization of the laws of the Member States with regard to the transport of dangerous goods by rail
- Council Directive 94/55/EC of 21 November 1994 on the harmonization of the laws of the Member States with regard to the transport of dangerous goods by road
- Regulation (EC) No 2099/2002 of the European Parliament and of the Council of 5 November 2002 establishing a Committee on Safe Seas and the Prevention of Pollution from Ships (COSS) and amending the Regulations on maritime safety and the prevention of pollution from ships
- Council Directive 96/35/EC of 3 June 1996 on the appointment and vocational qualification of safety advisers for the transport of dangerous goods by road, rail and inland waterways

International Treaties

GHS – Globally Harmonized System of Classification and Labelling of Chemicals

STATISTICS

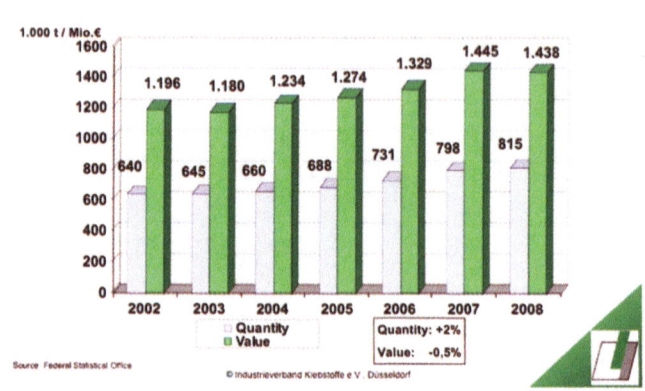

Adhesives Production in Germany

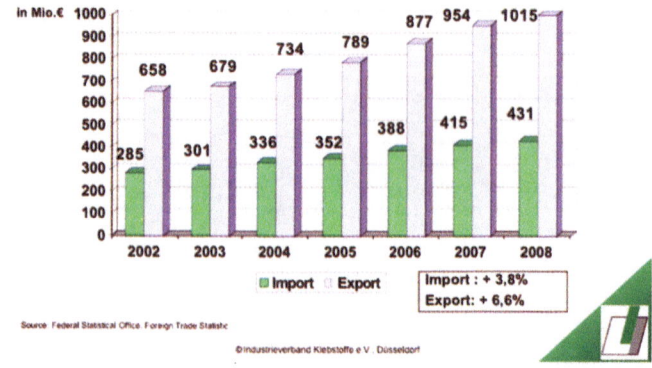

Imports and Exports of Adhesives

Development of selected Industries in Germany
Year on Year Growth (%)

Germany	%	2008	Forcast 2009	2009 / 01-03
Means of Transport	14,1	-3,0	-15,0	-33,8
Food, Beverage & Tobacco	8,7	-1,2	-2,1	-1,9
Paper / Print	7,4	-1,1	-5,2	-7,4
Metals & Metal Products	13,0	0,7	-11,0	-30,6
Plant and Machinery	14,8	4,2	-12,0	-17,9
Electrical & optical Equipment	15,4	4,2	-11,1	-20,3
Chemistry	9,7	-2,7	-9,4	-18,4
Wood & Furniture	3,5	-2,1	-7,0	-0,7
Main Contruction Industry		-0,7	-7,9	-12,3

Source: FERI; The Economist

© industrieverband Klebstoffe e.V., Düsseldorf

Growth Forecast for Germany 2009
GDP Growth (%)

Real GDP for Germany
Development adjusted by Season and Calendar Month

1) Growth compared with the previous quarter (%), annualized (scale on the right)
2) Numbers: Change in original values compared with the previous year

GD spring 2009

Source: Federal Statistical Office, Calculations and forecasts by institutes

© Industrieverband Klebstoffe e.V., Düsseldorf

General Economic Data
Growth over the previous Quarter (%), adjusted by Season)

	I/08	II/08	III/08	IV 08	I 09
GDP	1,5	-0,5	-0,5	-2,2	-3,8
Capital Investment	3,4	-1,4	0,2	-2,7	-7,9
Construction Investment	5,5	-3,4	-0,1	-1,3	-2,6
Private Consumption	-0,3	-0,6	0,3	-0,1	-1,7
Public Consumption	1,0	0,5	0,1	-0,1	0,3
Export - total	2,5	-0,3	-0,4	-8,1	-9,7
Import - total	3,2	-1,8	4,0	-4,1	-5,4

Adjusted by season and calender month revised by the Federal Statistical Office

Source: Federal Statistical Office

© Industrieverband Klebstoffe e.V., Düsseldorf

German Car Market 2009
(Passenger Cars/Estates)

Germany (Mio. Einheiten)	2007	2008	Δ %	March 2009	Δ %	1. Quarter 2009	Δ %	Jan. - April 2009	Δ %
Production	5,71	5,53	-3,1	0,34	-20,0	1,00	-33	1,43	-31
Exports	4,30	4,13	-4,0	0,27	-25,0	0,71	-38	0,94	-40
Domestic Registrations	3,15	3,09	-1,8	0,40	40,0	0,87	18	0,25	18
- German Brands	2,19	2,16	-1,3	0,24	27,0	0,54	9	n/a	n/a
-Foreign Brands	1,14	1,10	-3,6	0,16	65,0	0,32	36	n/a	n/a

Source: VDA

© Industrieverband Klebstoffe e.V., Düsseldorf

EMU – Industrial Production
(Y/Y Percentage Change)

Industrial Sectors	2007 Actual	2008 Forecast	2009 Forecast	2010 Forecast
Manufacturing	4,2	-1,9	-10,9	3,2
Transport	4,3	-4,5	-19,4	9,3
Electronics/ Computer	6,9	1,9	-11,3	3
Paper & Paper Products/Printing	2,3	-3,1	-8,4	1,6
Food & Bev. & Tobacco	1,6	-1,4	-1,7	0,3

© Industrieverband Klebstoffe e.V . Düsseldorf

Global Economic Development
(Y/Y Change in %)

Country	2007 Actual	2008 Forecast	2009 Forecast
USA	1,4	-3,2	-12,9
Mexico	2,6	-0,4	-4,7
Japan	3,0	-3,3	-22,9
China	19,8	12,9	7,8
Korea	6,9	3,1	-15,1
EMU	4,2	-1,9	-10,9
Germany	6,9	0,2	-10,4
France	1,9	-3,1	-10,3
Italy	2,3	-3,3	-12,6
Spain	2,1	-8	-14,1
UK	0,2	-2,7	-9,5
Turkey	6,7	-1,9	-11,2

Quelle: FERI Research The Economist   © Industrieverband Klebstoffe e.V . Düsseldorf

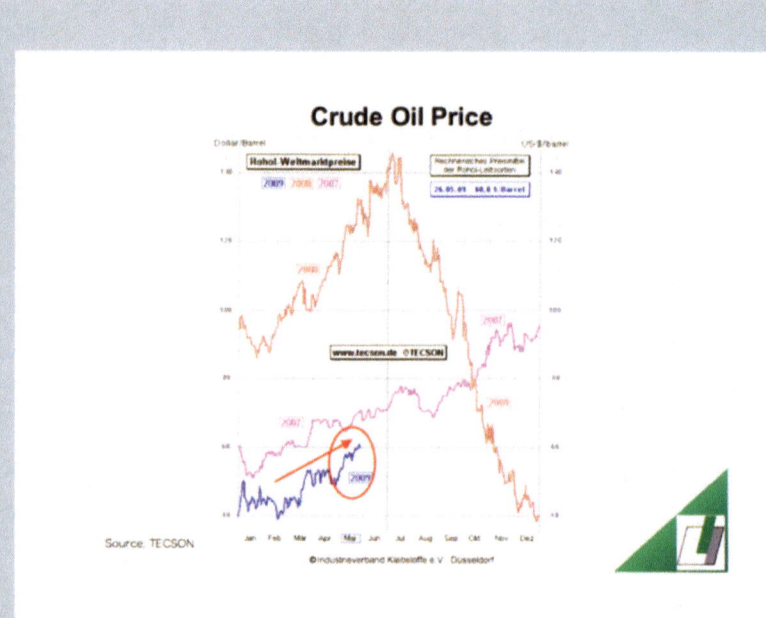

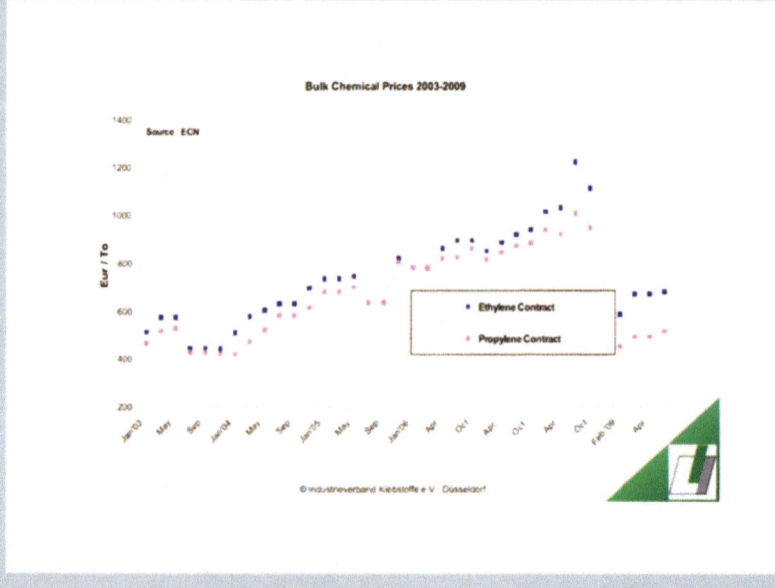

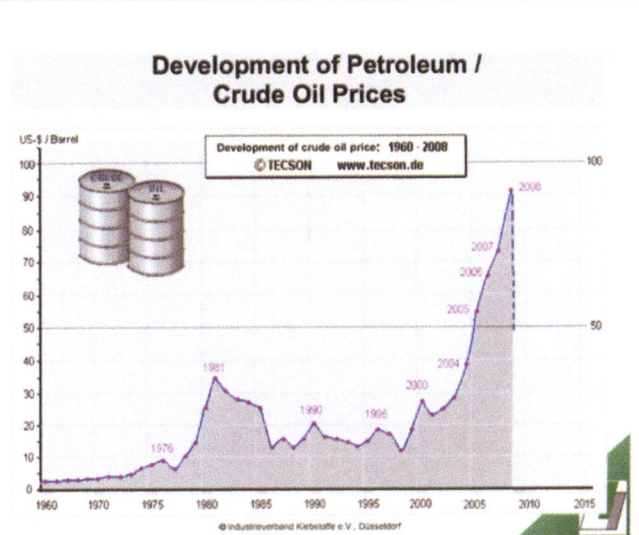

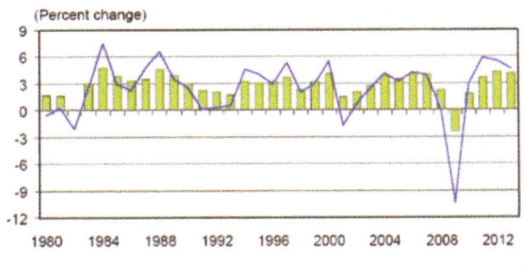

STANDARDS

Adhesives Standards

Early in the 1950s, adhesives standardization started in Germany at a national level. This initiative continued with the issuing of a terminology standard and other standards for various fields (wood, shoes, flooring adhesives). DIN standards issued during that period by the German Institute for Standardization (DIN) still apply on a domestic level unless they have been repealed or replaced by European standards. In 1962, the "COMITÉ EUROPÉEN DE NORMALISATION" (European Committee for Standardization, CEN) was founded. Thereafter, CEN established various technical committees (Technical Committee 52 "Safety of toys", Technical Committee 67 "Adhesives for tiles", Technical Committee 193 "Adhesives", Technical Committee 253 "Self adhesive tapes", Technical Committee 261 "Health informatics" and Technical Committee 264 "Indoor air quality") to establish and revise standards in the field of adhesives. Thus for the first time the opportunity was created to develop a broad set of standards applicable throughout Europe, and which was specifically aimed at adhesives manufacturers and consumers. This brought with it the chance for them to better represent their own economic and technical interests. Agreements which CEN signed with the "INTERNATIONAL ORGANIZATION FOR STANDARDIZATION" (ISO) meant that the global ISO standards were accepted as European standards and the two organizations went on to jointly work out and issue additional standards.

European standards are generally publicized in the three official languages of the European Union, i.e., German, English and French, and conform with one another from a technical perspective. European standardization supports trade across the EU by reducing trade barriers and is part of the framework of efforts to achieve technical harmonization. European standardization is supported by the German Adhesives Association. Unlike other standards, e.g., international ISO standards, European standards (EN) are binding at a European level. Both the European Court of Justice and national courts within the EU are obliged to base their jurisdiction on European standards.

The listed standards are sorted by areas, i.e.,
▸ Terminology, general physical and chemical test procedures
▸ Structural adhesives
▸ Adhesives for paper, board, packaging and disposable sanitary products
▸ Building adhesives
▸ Adhesives for leather and footwear materials
▸ Adhesives for thermoplastic piping systems
▸ Adhesives for wood and related timber products
▸ Adhesives for tiles
▸ Self adhesive tapes
▸ DIY, Consumer, Stationary
▸ Indoor air quality

Source for all quoted standards: Beuth-Verlag, Berlin (www.beuth.de).

Standards issued by DIN which only bear the designation "DIN ..." apply exclusively to Germany. Standards named "DIN EN ..." are European standards which DIN in its function as a member organization of CEN has accepted and publicized in a German version. If these have the same content as global ISO standards, they are publicized under the name "DIN EN ISO ...". European pre-standards are marked "DIN V ENV ...".

The European standards listed were developed by Technical Committees 52, 67, 193, 253, 261 and 264 of CEN (in the directories below this information is detailed in brackets after the headings). Information on newly issued European standards, titles and status of standardization projects and on European standards which are currently undergoing the mandatory 5 year review can be obtained from the above committees on the internet at http://www.cen.eu/CENORM/Sectors/Technical Committees and Workshops/List of technical Committees/Workshops and other bodies/Select Technical body/go/, from current directories /"Published Standards" and /"Standards under development" respectively.

Terminology, General Physical and chemical Test Procedures (CEN/TC193/WG1)

European Standards

DIN EN 542:2003
"Adhesives. Determination of Density"

DIN EN 543:2003
Determination of apparent density of powder and granule adhesives

DIN EN 827:2005
Adhesives. Determination of conventional solids content and constant mass solids content

DIN EN 923:2005
Adhesives. Terms and definitions

DIN EN 924:2003
Adhesives. Solvent-borne and solvent-free adhesives. Determination of flashpoint

DIN EN 1067:2005
Adhesives. Examination and preparation of samples for testing

DIN EN 1238:1999
Adhesives. Determination of the softening point of thermoplastic adhesives (ring and ball)

DIN EN 1239:1998
Adhesives. Freeze-thaw stability

DIN EN 1240:1998
Adhesives. Determination of hydroxyl value and/or hydroxyl content

DIN 1241:1998
Adhesives. Determination of acid value

DIN EN 1242:2005
Adhesives. Determination of isocyanate content

DIN EN 1243:1998
Adhesives. Determination of free formaldehyde content in amino and amino formaldehyde condensates

DIN EN 1244:1998
Determination of the colour and/or colour changes of adhesive coats under the influence of light

DIN EN 1245:1998
Adhesives. Determination of pH. Test method

DIN EN 1246:1998
Adhesives. Determination of ash and sulfated ash

DIN EN 113999
Adhesives. Short-term methods for measuring the emission properties of low solvent or solvent-free adhesives after application
Part 1: 2006 "General method"
Part 2: 2007 "Determination of volatile organic substances"
Part 3: 2007 "Determination of volatile aldehydes"
Part 4: 2007 "Determination of volatile di-isocyanates"

DIN EN 12092:2001
Adhesives. Determination of viscosity

DIN EN 12962:2001
Adhesives. Determination of elastic behaviour of liquid adhesives (elasticity index)

DIN EN 12963:2001
Adhesives. Determination of free monomer content in adhesives based on synthetic polymers

International Standards
DIN EN ISO 9665:2000
Adhesives. Animal glues. Methods of sampling and testing

DIN EN ISO 10363: 1995
Hot-melt adhesives. Determination of thermal stability

DIN EN ISO 14678:2005
Adhesives. Determination of resistance to flow

DIN EN ISO 15605: 2004
Adhesives. Sampling

Structural Adhesives (CEN/TC193/WG2)

DIN 53287:1979
Testing of metal adhesives and metal bonds. Determining resistance to liquids

DIN 54455:1984
Testing of metal adhesives and metal bonds. Torsion shear test.

DIN 54456:1989
Testing of constructive bonds. Weather resistance test

European standards

DIN EN 828:1997
Adhesives - Wettability - Determination by measurement of contact angle and critical surface tension of solid surface

DIN EN 1464:1994
Adhesives. Determination of peel resistance of high-strength adhesive bonds - Floating roller method

DIN EN 1465:1994
Adhesives. Determination of tensile lap-shear strength of rigid-to-rigid bonded assemblies

DIN EN 1965
Structural adhesives. Corrosion
Part 1: Determination and Classification of Corrosion to a Copper Substrate
Part 2: Determination and Classification of Corrosion to a Brass Substrate

DIN EN 1966:2002
Structural adhesives. Characterisation of a surface by measuring adhesion by means of the three point bending method

DIN EN 1967:2002
Structural adhesives. Evaluation of the effectiveness of surface treatment techniques for aluminium using a wet peel test in association with the floating roller method

DIN EN 12701:2001
Structural adhesives. Storage. Determination of words and phrases relating to the product life of structural adhesives and related materials

DIN EN 13887:2003
Structural adhesives. Guidelines for surface preparation of metals and plastics prior to adhesive bonding

DIN EN 14022:2003
Structural adhesives. Determination of the pot life (working life) of multicomponent adhesives

DIN EN 14258:2004
Structural adhesives. Mechanical behaviour of bonded joints subjected to short and long terms exposure at specified conditions of temperature

DIN EN 14444:2005
Structural adhesives. Qualitative assessment of durability of bonded assemblies. Wedge rupture test

DIN EN 14896
Structural adhesives. Determination of shear behaviour of structural bonds
Part 1: 2004 Torsion test method using butt-bonded hollow cylinders
Part 2: 2004 Thick adherends shear test

DIN EN 15190:2007
Structural adhesives. Test methods for assessing long term durability of bonded metallic structures

DIN EN 15274:2007
General purpose adhesives for structural assembly - Requirements and test methods

DIN EN 15275:2007
Structural adhesives - Characterisation of anaerobic adhesives for co-axial metallic assembly in building and civil engineering

DIN EN 15336:2007
Adhesives. Determination of the time to rupture of bonded joints under static load

DIN EN 15337:2007
Adhesives. Determination of shear of anaerobic adhesives using pin-and-collar specimens

DIN EN 26922:1993
Adhesives. Determination of tensile strength of butt joints

DIN EN 28510:1993
Adhesives. Peel test for a flexible-bonded-to-rigid test specimen
Part 1:1993 90° peel test
Part 2:1993 180° peel test

CEN/TR 14548:2003
Adhesives. Guide to test methods and other standards for the general requirements, characterization and safety of structural adhesives

International standards

DIN EN ISO 9142:2003
Adhesives. Guide to the selection of standard laboratory ageing conditions for testing bonded joints

DIN EN ISO 9653:2000
Adhesives. Test method for shear strength of adhesive bonds

DIN EN ISO 9664:1995
Adhesives. Test method for fatigue properties odf structural adhesives in tensile shear

DIN EN ISO 10365:1995
Adhesives. Designation of main failure patterns

DIN EN ISO 10964:1997
Adhesives. Determination of torque strength of anaerobic adhesives on threaded fasteners

DIN EN ISO 11339:2005
Adhesives. T-peel test for flexible-to-flexible bonded assemblies

DIN EN ISO 11343:2005
Adhesives. Determination of dynamic resistance to cleavage of high strength adhesive bonds under impact conditions. Wedge impact method

DIN EN ISO 13445:2006
Adhesives. Determination of shear strength of adhesive bonds between rigid substrates by the block-shear method

European standards

DIN EN 17:19 1998
Adhesives for paper and board, packaging and disposable sanitary products. Tack measurement for pressure-sensitive adhesives. Determination of loop tack

DIN EN 1720:1998
Adhesives for paper and board, packaging and disposable sanitary products. Determination of dispersability

DIN EN 1721:1998
Adhesives for paper and board, packaging and disposable sanitary products. Tack measurement for pressure sensitive adhesives. Determination of rolling ball tack

DIN EN 1895:2001
Adhesives for paper and board, packaging and disposable sanitary products. – "T" peel test for a flexible-to-flexible assembly

DIN EN 1896:2001
Adhesives for paper and board, packaging and disposable sanitary products. Determination of tensile strength and elongation

DIN EN 12702:2002
Adhesives for paper and board, packaging and disposable sanitary products. Determination of blocking behaviour of potentially adhesive layers

DIN EN 12703:1999
Adhesives for paper and board, packaging and disposable sanitary products. Determination of low temperature flexibility or cold crack temperature

DIN EN 12704:1999
Adhesives for paper and board, packaging and disposable sanitary products. Determination of foam formation for aqueous adhesives

DIN EN 12960:2001
Adhesives for paper and board, packaging and disposable sanitary products. Determination of shear resistance

DIN EN 14713:2005
Adhesives for paper and board, packaging and disposable sanitary products. Determination of friction properties of films potentially suitable for bonding

CR 14376:2002
Adhesives for paper and board, packaging and disposable sanitary products. Description and assessment of the setting process

Building Adhesives (CEN/TC193/WG1)

European standards
DIN 1372:1999
Adhesives. Test method for adhesives for floor and wall coverings. Peel test

DIN EN 1373:1999
Adhesives. Test method for adhesives for floor and wall coverings. Shear test

DIN EN 1841:1998
Adhesives. Test methods for floor coverings and wall coverings. Determination of dimensional changes of a linoleum floor covering in contact with an adhesive

DIN EN 1902:1999
Adhesives. Test methods for adhesives for floor coverings and wall coverings. Shear test

DIN 1993:1999
Adhesives. Test method for adhesives for plastic or rubber floor coverings or wall coverings. Determination of dimensional changes after accelerated ageing

DIN EN 1937:1999
Test method for hydraulic setting floor smoothing and/or levelling compounds. Standard mixing procedures

DIN EN 12706:1999
Test methods for hydraulic setting floor smoothing and/or levelling compounds. Determinatiion of flow characteristics

DIN EN 13408:2002
Methods of test for hydraulic setting floor smoothing and/or levelling compounds. Determination of bond strength

DIN EN13409:2002
Methods of test for hydraulic setting floor smoothing and/or levelling compounds. Determination of setting time

DIN EN 13415:2002
Adhesives. Test of adhesives for floor coverings - Determination of the electrical resistance of adhesive films

DIN EN 13851:2003
Methods of test for hydraulic setting floor smoothing and/or levelling compounds. Determination of flexural and compressive strength

DIN EN 13872:2004
Method of test for smoothing and/or levelling compounds. Determination of dimensional change

DIN EN 14259:2003
Adhesives for floor covering. Requirements for mechanical and electrical performance

DIN EN 14293:2006
Adhesives. Adhesives for bonding parquet to subfloor. Test methods and minimum requirements.

Adhesives for Leather and Footwear Materials (CEN/TC 193 WG5)

DIN EN 1392:2006
Adhesives for leather and footwear materials. Solvent-based and dispersion adhesives. Test methods for measuring the bond strength under specified conditions

DIN EN 12705:1999
Adhesives for leather and footwear materials. Determination of colour change of white or bright coloured leather surfaces by migration

DIN EN 12961:2001
Adhesives for leather and footwear materials. Determination of optimum activation temperatures and maximum activation life of solvent-based and dispersion adhesives

DIN EN 12964:2001
Adhesives for leather and footwear materials. Lasting adhesives. Testing heat resistance of bonds at increasing temperature

DIN EN 14294:2004
Adhesives for leather and footwear materials. Preparation of bonded test pieces by moulding-on processes

DIN EN 14510:2005
Adhesives for leather and footwear materials. Solvent-based and dispersion adhesives. Determination of sole positioning tack (spotting tack)

DIN EN 15062:2007
Adhesives for leather and footwear materials. Solvent-based and dispersion adhesives. Testing ageing of bonds under specified conditions

DIN EN 15307:2007
Adhesives for leather and footwear materials. Sole-upper bonds. Minimum strength requirements

Adhesives for thermoplastic Piping Systems (CEN/TC193/SC1)

European standards
DIN EN 14680:2006
Adhesives for non-pressure thermoplastic piping systems. Specifications

DIN EN 14814:2006
Adhesives for thermoplastic piping systems for fluids under pressure. Specification.

DIN CEN/TS 14999:2003
Adhesives for thermoplastic piping systems. Test of adhesives under artificial ageing

International standards
DIN EN ISO 9311
Adhesives for thermoplastic piping systems.
Part 1:2005 Test method for the determination of thermal stability of adhesives
Part 2:2005 Determination od shear strength
Part 3:2005 Test method for the determination of resistance to internal pressure

DIN EN ISO 15908:2002
Adhesives for thermoplastic piping systems. Test method for the determination of thermal stanbility of adhesives

Adhesives for Wood and Related Timber Products (CEN/TC193SC1)

European standards
DIN EN 204:2001
Classification of thermoplastic wood adhesives for non-structural applications

DIN EN 205:2003
Adhesives. Wood adhesives for non-structural applications. Determination of tensile shear strength of lap joints

DIN EN 301:2006
Adhesives, phenolic and aminoplastic, for load bearing timber structures. Classification and performance requirements

DIN EN 302
Adhesives for load bearing timber structures. Test methods
Part 1:2004 Determination of bond strength in longitudinal tensile shear strength
Part 2:2004 Determination of resistance to delamination
Part 3:2004 Determination of the effect of acid damage to wood fibres by temperature and humidity cycling on the transverse tensile strength
Part 4:2004 Determination of the effects of wood shrinkage on the shear strength
Part 5:ENV:2001 Determination of the conventional assembly time
Part 6:2004 Determination of the conventional pressing time
Part 7:2004 Determination of the conventional working life

DIN EN 12436:2001
Adhesives for load-bearing timber structures. Casein adhesives. Classification and performance requirements

DIN EN 12765:2001
Classification of thermosetting wood adhesives for non-structural applications

DIN EN 14256:2007
Adhesives for non-structural wood applications. Test method and requirements for resistance to static load

DIN EN 14257:2006
Adhesives. Wood adhesives. Determination of tensile strength of lap joints at elevated temperature

DIN EN 14292:2005
Adhesives. Wood adhesives. Determination of static load resistance with increasing temperature (WATT 91)

DIN EN 15416
Adhesives for load bearing timber structures other than phenolic and aminoplastic

Part 1:2006 Static load test of single bondline specimens in compression shear
Part 2:2006 Static load test of multiple bondline specimens in compression shear
Part 3:2008 Creep deformation test at cyclic climate conditions with specimens loaded in bending shear
Part 4:2006 Test methods. Determination of open assembly time for one component polyurethane adhesives
Part 5:2006 Determination of conventional pressing time

Din EN 15425:2006
Adhesives. One component polyurethane for load bearing timber structures. Classification and performance requirements

Adhesives for Tiles (CEN/TC67)

European standards
DIN EN 1308:1996 (:1998)
Adhesives for tiles. Determination of slip

DIN EN 1323:2007
Adhesives for tiles. Concrete slabs for tests
DIN EN 1324:2007
Adhesives for tiles. Determination of shear adhesion strength of dispersion adhesives.

DIN EN 1346:2007
Adhesives for tiles. Determination of open time

DIN EN 1347:2007
Adhesives for tiles. Determination of wetting capability

DIN EN 1348:1997 (:1998)
Adhesives for tiles. Determination of tensile adhesion strength for cementitious adhesives

DIN EN 12002:2002
Adhesives for tiles. Determination of transverse deformation for cementitious adhesives and grouts

DIN EN 12003:2008
Adhesives for tiles. Determination of shear adhesion strength of reaction resin adhesives

DIN EN 12004:2007
Adhesives for tiles. Definitions and specifications

DIN EN 12008
Part 1:2008 Adhesives and grouting for tiles. Determining the resistance of reaction resin
grouting mortar to chemicals
Part 2:2008 Grouting for tiles. Determining abrasion resistance
Part 3:2008 Grouting for tiles. Determining flexural and compressive strength
Part 4:2008 Grouting for tiles. Determining shrinkage
Part 5:2008 Grouting for tiles. Determining water absorption

DIN EN 13888:2002
Grouting for tiles. Definitions and specifications

DIN EN 14891:2007
Liquid applied water impermeable products for use beneath ceramic tiling bonded with adhesives.
Requirements, test methods, evaluation of conformity, classification and designation

Self Adhesive Tapes (CEN/TC253)

European standards
DIN EN 1939:2003
Self adhesive tapes – Detremination of peel adhesion properties

DIN EN 1942:2003
Self adhesive tapes. Measurement of thickness

DIN EN 1943:2003
Self adhesive tapes. Measurement of static shear adhesion

DIN 1944:1996
Self adhesive tapes. Measurement of unwind adhesion at low speed
DIN EN 1945:1996
Self adhesive tapes. Measurement of quick stick

DIN EN 12023:1996
Self adhesive tapes. Measurement of water vapour transmission in a warm humid atmosphere

DIN EN 12024:1996
Self adhesive tapes. Measurement of resistance to elevated temperature and humidity

DIN EN 12025:1996
Self adhesive tapes. Measurement of tear resistance by the pendulum method

DIN EN 12026:1996
Self adhesive tapes. Measurement of the unwinding force at high speed

DIN EN 12027:1996
Self adhesive tapes. Measurement of flame resistance

DIN EN 12028:1996
Self adhesive tapes. Measurement of elongation under static load

DIN EN 12029:1996
Self adhesive tapes. Determination of the water-soluble corrosive ions

DIN EN 12030:1996
Self adhesive tapes. Measurement of impact resistance

DIN EN 12031:1996
Self adhesive tapes. Measurement of bursting strength

DIN EN 12032:1996
Self adhesive tapes. Measurement of bonding of thermosetting adhesive tapes during curing

DIN EN 12033:1996
Self adhesive tapes. Measurement of bonding of thermosetting adhesive tapes after curing

DIN EN 12034:1996
Self adhesive tapes. Measurement of the length of a roll of adhesive tape

DIN EN 12035:1996
Self adhesive tapes. Flagging of adhesive tape

DIN EN 12036:1996
Self adhesive tapes. Solvent penetration into adhesive masking tapes

DIN EN 12481:2000
Self adhesive tapes. Terminology

DIN EN 14410:2003
Self adhesive tapes. Measurement of breaking strength and elongation at break

Household, DIY, Consumer, Stationary (CEN/TC52 and CEN/TC261)

European standards

DIN EN 71
Safety of toys
Part 1:2005 Mechanical and physical properties (:2006)
Part 2:2006 Flammability
Part 3:2003 Specification for migration of certain elements (:2003)
Part 4:1990 Experimental sets for chemistry and related activities
Part 5:1993 Chemical toys (sets) other than experimental sets (:2006)
Part 6:1999 Graphical symbols for age warning labelling
Part 7:2000 Finger paints. Requirements and test methods
Part 8:2003 Swings, slides and similar activity toys for indoor and outdoor family domestic use (:2005)
Part 9:2005 Organic chemical compounds. Requirement
Part 10:2005 Organic chemical compounds. Sample preparation and extraction
Part 11:2005 Organic chemical compounds. Methods of analysis

DIN EN ISO 8317:2004
Child-resistant packaging. Requirements and testing procedures for reclosable packages

DIN EN 862:2005
Packaging. Child resistant packaging. Requirements and testing procedures for non-reclosable packages for non-pharmaceutical products

DIN EN 13427:2004
Packaging. Requirements for the use of European Standards in the field of packaging and packaging waste

DIN 13428:2004
Packaging. Requirements specific to manufacturing and composition. Prevention by source reduction

DIN 13429:2004
Packaging. Reuse

DIN 13430:2004
Packaging. Requirements for packaging recoverable by material recycling
DIN 13431:2004

Packaging. Requirements for packaging recoverable in the form of energy recovery, including specification of minimum inferior calorific value

DIN 15178:2007
Elements for the identification of products in emergency enquiries

International standards
DIN EN ISO 10993:1997
Biological evaluation of medical devices. Evaluation and testing

DIN EN ISO 11683:1997
Packaging. Tactile warnings of danger. Requirements

Indoor Air Quality (CEN/TC264)

International standards
DIN EN ISO 16000-1:2006
Indoor air. General aspects of sampling strategy

DIN EN ISO 16000-2:2006
Indoor air. Sampling strategy for formaldehyde

DIN EN ISO 16000-3:2002
Indoor air. Measurement of formaldehyde and other carbonyls in air by pumped sampling

DIN EN ISO 16000-4:2002
Indoor air. Diffusive sampling of formaldehyde in air

DIN EN ISO 16000-5:2007
Indoor air. Sampling strategy for volatile organic compounds (VOCs)

DIN EN ISO 16000-6:2004
Indoor air. Determination of the emission of volatile organic compounds from building products and furnishing. Emission test chamber method

DIN EN ISO 16000-9:2006
Indoor air. Determination of the emission of volatile organic compounds from building products and furnishing. Emission test chamber method

DIN EN ISO 16000-10:2006
Indoor air. Determination of the emission of volatile organic compounds from building products and furnishing. Emission test cell method

DIN EN ISO 16000-11:2006
Indoor air. Determination of the emission of volatile organic compounds from building products and furnishing. Sampling, storage of samples and preparation of test specimens

SOURCES

- ▸ Raw Materials
- ▸ Adhesives by Types
- ▸ Sealants
- ▸ Adhesives by Key Market Segments
- ▸ Equipment
- ▸ Technical Consultancy

Raw materials

AB-Tec
Alberdingk Boley
BASF
Bayer MaterialScience
ChemCycle Bitterfeld
Celanese
DKSH
Evonik
ExxonMobil Chemical
Hanse Chemie
HexionSpeciality
IMCD
KANEKA Belgium
Krahn Chemie
Kraton Polymers
LANXESS
MORCHEM
Newport
Omya Peralta
ORGANIK KIMYA
Polimeri Europa
Polymer Latex
Rütgers
Sonderhoff
Synthopol Chemie
Wacker Chemie
Worlée-Chemie

Adhesives by Types
Hot-melt Adhesives

AB-Tec
Bostik
Friedrich Branding
BÜHNEN
ChemCycle Bitterfeld
DKSH
Dow Automotive
Eluid Adhesive
EUKALIN
Forbo Adhesives
Forbo Erfurt
H.B. Fuller

GLUDAN
Henkel
Huntsman
Jowat
Klebstoffwerke COLLODIN
Kleiberit
Kömmerling
MORCHEM
Newport
Novamelt-Jowat
PLANATOL Adhesive
PRHO-CHEM
Rhenocoll
SABA Dinxperlo
Sika Automotive
Türmerleim
UHU
Vito Irmen
Weiss Chemie + Technik
Wetzel
ZELU CHEMIE

Reactive Adhesives

ARDEX
Bona
Bostik
BYLA
CHEMETALL
COIM Deutschland
Cyberbond
DEKA
DELO
DKSH
Dow Automotive
Dymax Europe
fischer
Forbo Erfurt
H.B. Fuller
Henkel
Huntsman
Jowat
Kiesel Bauchemie
Kisling
Kleiberit
Kömmerling

LUGATO CHEMIE
merz + benteli
Mlnova
MORCHEM
Otto-Chemie
Panacol-Elosol
PCI
Planatol Adhesive
SABA Dinxperlo
Schomburg
SCHÖNOX
Sika Automotive
Sika Deutschland
Sonderhoff
STAUF
Uzin Utz
Wakol-Chemie
Weiss Chemie + Technik
ZELU CHEMIE

Dispersion Adhesives
Bona
Bostik
Friedrich Branding
DEKA
Eluid Adhesive
EUKALIN
fischer
Follmann
Forbo Adhesives
Forbo Erfurt
H.B. Fuller
GLUDAN
Henkel
IMCD
Jowat
Kiesel Bauchemie
Klebstoffwerk COLLODIN
Kleiberit
Kömmerling
LUGATO CHEMIE
Newport
ORGANIK KIMYA
PCI
PLANATOL Adhesive

PRHO-CHEM
Renia-Gesellschaft
Rhenocoll
SCHÖNOX
Schomburg
Sika Automotive
STAUF
Synthopol Chemie
Türmerleim
Vito Irmen
Wakol-Chemie
Weiss Chemie & Technik
Wetzel
ZELU CHEMIE

Vegetable Adhesives,
Dextrin and Starch Adhesives
Friedrich Branding
DKSH
Eluid
EUKALIN
Forbo Adhesives
H.B. Fuller
Henkel
Klebstoffwerke COLLODIN
PLANATOL Adhesive
PRHO-CHEM
SCHÖNOX
Türmerleim

Animal Glue
H.B. Fuller
Henkel
PRHO-CHEM

Solvent-based Adhesives
Adtracon
Bona
Bostik
ChemCycle Bitterfeld
COIM Deutschland
DEKA
DKSH
fischer
Forbo Adhesives

Forbo Erfurt
H.B. Fuller
IMCD
Jowat
Kiesel Bauchemie
Kleiberit
Kömmerling
LANXESS
Newport
Otto-Chemie
PLANATOL Adhesive
Renia-Gesellschaft
Rhenocoll
SABA Dinxperlo
SCHÖNOX
Sika Automotive
STAUF
Synthopol Chemie
Vito Irmen
Wakol-Chemie
Weiss Chemie + Technik
ZELU CHEMIE

Pressure-Sensitive Adhesives

Bostik
ChemCycle Bitterfeld
DEKA
DKSH
Dymax Europe
Eluid Adhesive
EUKALIN
Forbo Adhesives
H.B. Fuller
GLUDAN
Henkel
IMCD
Klebstoffwerke COLLODIN
Kleiberit
LANXESS
Novamelt-Jowat
ORGANIK KIMYA
PLANATOL Adhesive
PRHO-CHEM
Rhenocoll
Sika Automotive

Synthopol Chemie
Türmerleim
Vito Irmen
Wakol
Weiss Chemie + Technik
Wetzel
ZELU CHEMIE

Sealants

Berger-Seidle Siegeltechnik
Bostik
fischer
Forbo Adhesives
Hanno-Werk
Henkel
Kisling
merz + benteli
MINOVA
Newport
ORGANIK KIMYA
OTTO-Chemie
PCI
Schomburg
Sonderhoff
Synthopol Chemie

Adhesives by Key Market Segments

Self Adhesive Tapes

AB-Tec
certoplast Vorwerk & Sohn
Coroplast
Eluid Adhesive
IMCD
LANXESS
Lohmann
Novamelt-Jowat
PLANATOL Adhesive
Synthopol Chemie
tesa

Paper/Packaging
AB-Tec
Bostik
Friedrich Branding
certoplast Vorwerk & Sohn
ChemCycle Bitterfeld
COIM Deutschland
Coroplast
DKSH
DEKA
Eluid Adhesive
EUKALIN
Evonik
Forbo Adhesive
Forbo Erfurt
H.B. Fuller
GLUDAN
Henkel
IMCD
Jowat
Klebstoffwerke COLLODIN
LANXESS
Lohmann
MORCHEM
Novamelt-Jowat
Omya Peralta
ORGANIK KIMYA
PLANATOL Adhesive
PRHO-CHEM
Rhenocoll
Sika Automotive
Sonderhoff
Synthopol Chemie
tesa
Türmerleim
Wakol-Chemie
Weiss Chemie + Technik
Wetzel

Bookbinding/
Graphics Industry
Friedrich Branding
Eluid Adhesive
EUKALIN
Forbo Adhesives

H. B. Fuller
GLUDAN
Henkel
IMCD
Jowat
LANXESS
Lohmann
MORCHEM
National Adhesives
Omya Peralta
ORGANIK KIMYA
PLANATOL Adhesive
PRHO-CHEM
Sika Automotive
tesa
Türmerleim
Wetzel

Wood/Furniture industry
AB-Tec
Adtracon
Berger-Seidle Siegeltechnik
Bostik
Friedrich Branding
BÜHNEN
BYLA
Coroplast
Cyberbond
DEKA
DKSH
Eluid Adhesive
Evonik
fischer
Forbo Adhesives
Follmann
H.B. Fuller
Henkel
Jowat
Kisling
Kleiberit
Kömmerling
LANXESS
Lohmann
merz + benteli
MINOVA

Omya Peralta
ORGANIK KIMYA
Otto-Chemie
Panacol-Elosol
Rhenocoll
SABA Dinxperlo
Sika Automotive
STAUF
Synthopol Chemie
tesa
Türmerleim
Vito Irmen
Wakol-Chemie
Weiss Chemie + Technik
ZELU CHEMIE

Building and Construction Industry
including Floors, Walls and Ceilings

AB-Tec
ARDEX
Ashland Drew Ameroid Deutschland
Berger-Seidle Siegeltechnik
Bona
Bostik
Friedrich Branding
BYLA
certoplast Vorwerk & Sohn
ChemCycle Bitterfeld
Coroplast
DEKA
DKSH
fischer
Forbo Adhesives
Forbo Erfurt
H. B. Fuller
GLUDAN
Henkel
IMCD
Kiesel Bauchemie
Kisling
Kleiberit
Kömmerling
Lohmann
LUGATO CHEMIE
merz + benteli

MINOVA
Novamelt-Jowat
Omy Peralta
ORGANIK KIMYA
Otto-Chemie
PLANATOL Adhesive
PCI
Rhenocoll
SCHÖNOX
Schomburg
Sika Automotive
STAUF
Synthopol Chemie
tesa
Uzin Utz
Wakol-Chemie
Weiss Chemie + Technik

Car and Aircraft Industries

AB-Tec
Adtracon
Friedrich Branding
BÜHNEN
BYLA
certoplast Vorwerk & Sohn
CHEMETALL
Coroplast
Cyberbond
DEKA
DELO
DKSH
Deutsche Hutchinson
Dow Automotive
Dymax Europe
Evonik
Forbo Adhesives
H.B. Fuller
Henkel
Huntsman
Kisling
Klebstoffwerke COLLODIN
Kleiberit
Kömmerling
Lohmann
merz + benteli

Novamelt-Jowat
Otto-Chemie
Panacol-Elosol
PLANATOL Adhesive
Sika Automotive
Sika Deutschland
Sonderhoff
Synthopol Chemie
tesa
Vito Irmen
Wakol-Chemie
Weiss Chemie + Technik
ZELU CHEMIE

Electronics
BÜHNEN
BYLA
certoplast Vorwerk & Sohn
CHEMETALL
Coroplast
Cyberbond
DELO
DKSH
Dymax Europe
Evonik
H.B. Fuller
Henkel
Huntsman
Kisling
Kömmerling
Lohmann
merz + benteli
Otto-Chemie
Panacol-Elosol
Sika Automotive
Sika Deutschland
tesa
Weiss Chemie + Technik

Sanitary Industry
Bostik
EMS-Chemie
H.B. Fuller
GLUDAN
Henkel

Jowat
Kisling
Kömmerling
LANXESS
Lohmann
Novamelt-Jowat
PRHO-CHEM
Sika Automotive
Türmerleim
Vito Irmen

Assembly, General Industry
AB-Tec
Adhesives Research
BYLA
certoplast Vorwerk & Sohn
CHEMETALL
Coroplast
Cyberbond
DEKA
DELO
Dow Automotive
Henkel
Huntsman
Kisling
Kleiberit
Kömmerling
Lohmann
merz + benteli
Novamelt-Jowat
Otto-Chemie
Panacol-Elosol
Renia-Gesellschaft
SABA Dinxperlo
Schomburg
Sonderhoff
Synthopol Chemie
tesa

Textile Industry
Adtracon
Bostik
Friedrich Branding
BÜHNEN
DEKA

EUKALIN
Evonik
Forbo Adhesives
H.B. Fuller
Henkel
Jowat
Kisling
Kleiberit
LANXESS
MORCHEM
Novamelt-Jowat
Omya Peralta
ORGANIK KIMYA
SABA Dinxperlo
Sika Automotive
Synthopol Chemie
tesa
Vito Irmen
Wakol-Chemie
ZELU CHEMIE

Self Adhesive Tapes, Labels
AB-Tec
Bostik
DKSH
EUKALIN
H.B. Fuller
Henkel
IMCD
Jowat
LANXESS
ORGANIK KIMYA
Novamelt-Jowat
PLANATOL Adhesive
PRHO-CHEM
STAUF
Sika Automotive
Synthopol Chemie
Türmerleim
Vito Irmen
Wetzel

Household, Hobby,
Offices, Stationery
certoplast Vorwerk & Sohn

Coroplast
Cyberbond
fischer
GLUDAN
Henkel
Kisling
LUGATO CHEMIE
Omya Peralta
Panacol-Elosol
Renia-Gesellschaft
Rhenocoll
tesa
Weiss Chemie + Technik
Footwear & Leather Industry
Adtracon
BÜHNEN
Cyberbond
H.B. Fuller
Henkel
Kömmerling
Renia-Gesellschaft
Sika Automotive
Wakol-Chemie
ZELU CHEMIE

Equipment
for Adhesive Handling, Mixing, Dosing
and Application
BÜHNEN
Inatec
Robatech
Scheugenpflug
Sulzer Mixpack
TIGRES

Surface Treatment
TIGRES

Technical consultancy
ChemQuest Europe
Hinterwaldner Consulting

We'll connect **You** to the top

Fax +49(0)5241 809620

 +

Yes, I would like to receive **adhesive** delivered to my address for one year **4 issues + free access to our online archive + free Adhesive Technology Compendium** for the special subscription price of € 93 plus postage. This price is guaranteed until the end of 2009. My subscription will be automatically extended for a further year unless I cancel at least 6 weeks before the end of the subscription period.

Name

Company

Industry

Position

Address

Country

e-mail

Date Signature   311 09 402

www.adhaesion.com

GWV Fachverlage GmbH | AG Wiesbaden HRB 9754 | Managing Directors: Dr. R. Birkelbach, A.F. Schirmacher